Classical Mechanics and Chaos

**Book 1 of Physics from Maximal Information Emanation,
a seven-book physics series.**

ISBN 979-8-9888160-1-0

Classical Mechanics and Chaos

by

Stephen Winters-Hilt

ISBN 979-8-9888160-1-0

Golden Tao Publishing
Angel Fire, NM
USA

Dedication

This book is dedicated to my family that helped on this lengthy road of discovery: Cindy, Nathaniel, Zachary, Sybil, Eric, Joshua, Teresa, Steffen, Hannah, Anders, Angelo, John and Susan.

Contents

Preface to Physics Series on:

Physics from Maximal Information Emanation

> "The Road goes ever on and on
> Down from the door where it began.
> Now far ahead the Road has gone,
> And I must follow, if I can,
> Pursuing it with eager feet,
> Until it joins some larger way
> Where many paths and errands meet.
> And whither then? I cannot say"

— J.R.R. Tolkien, The Fellowship of the Ring

Variation, Propagation, and Emanation

This is a seven book Physics Series that starts with Classical Mechanics (Book 1 [46]), then Classical Field Theory, such as electromagnetism (Book 2 [40]), then Manifold Dynamics, such a General Relativity (Book 3 [41]). The switch to a quantum mechanics description is given in Book 4 [42], and to a quantum field theory, QED in particular, in Book 5 [43]. A 'quantum manifold theory' would be the obvious next step except it cannot be done (there is not a renormalizable Field theory for Gravitation). Instead a thermal quantum manifold theory is considered, as well as Black Hole thermodynamics in general, in Book 6 [44]. Book 7 [45] describes a new theory, Emanator Theory, that provides a deeper mathematical construct that undergirds Quantum theory, much like quantum theory can be shown to provide a deeper (complexified) mathematical construct based on the classical theory.

This is a modern exposition where subtleties of chaos theory are described in Book 1, of Lorentz Invariance in Book 2, of Covariant Derivatives (General Relativity) and Gauge Covariant Derivatives (Yang-Mills Field Theory) in Book3. Book 4 on Quantum Mechanics provides an extensive review of QM, then considers a full self-adjoint analysis on the full general relativistic solution to the spherical shell in-fall system (a result carried over from Book 3). Book 5 considers QFT basics in detail, along with alternate vacua in specific scenarios. Book 6 considers thermodynamics from the basics to the Hamiltonian thermodynamics of some Black Hole systems. Throughout, the odd recurrence of the alpha parameter is noted. In Book 7 we look to a deeper mathematical formulation from which the Quantum Path Integral formulation would result, as well as explaining the odd parameters and structures that have been discovered (such as alpha and Lorentz Invariance).

The physical description starts with the classic formulations of point particle motion. The first approach to doing this is using differential equations (Newton's 1st and 2nd Law); the second is using a variational function formulation to select the differential equation (Lagrangian variation); the third is using a variational functional formulation (Action formulation) to select the variational function formulation. Historically, it wasn't realized until much later that there are two domains for motion in many systems: non-chaotic; and chaotic.

In a description of particle motion, assuming not in a parameter domain with chaotic motion, several important limits are found to exist. Examples include: the universal constants from the aforementioned chaos phenomenon, that are still encountered in non-chaos regimes if driven "to the edge of chaos". Limits are found where scattering is defined in the asymptotic limit and perturbation theory is well-defined in the sense that it is convergent. Overall, if the evolution is described as a 'process' it is often a Martingale process, which has well-defined limits. So, we have descriptions for motion, typically reducible to an ordinary differential equation (ODE), and for which solutions (requiring limit-definitions) are typically found to exist.

The physical description then contends with field dynamics in 2D, 3D, and 4D (in Book 3 [41]). Two-dimensional ("2D") field dynamics can be described as a complex function (that maps complex numbers to complex numbers). A novelty of the 2D complex function is it also shows how to handle many types of singularities (the residue theorem), thus provides important information about fundamental structures in physics as well as fundamental mathematical techniques for solving many integrals. For the 3D field dynamics we do an analysis of the electromagnetic field in 3D. The level of coverage begins at an overview of electrostatics at the level of the graduate text Jackson [123]. Some problems from Jackson Ch's 1-3 are examined closely in developing the theory itself. For some this material (in Book 2 [40]) might provide a useful accompaniment to Jackson's text in a full course on EM (based from Jackson's text). A quick review of electrodynamics and electromagnetic wave phenomena is then given. In essence, we see many more examples of ODE problems with solutions, such as for the 3D Laplacian, usually involving separation of variables. We then review the famous transform, discovered by Lorentz in 1899 [124], that relates the EM field as seen by two observers differing by a relative velocity. With the existence of this transform, that brings in the time dimension along with the relative velocity, we effectively have a 4D theory.

From Lorentz Invariance we have, as a point transformation, rotational invariance under SO(3) or SU(2). If Lorentz Invariance is fundamental, then we should see both forms of rotation invariance, one of vector/tensor type from SO(3), and one of spinorial type from SU(2). This is the case, as gauge fields are vectorial and matter fields are spinorial. From Lorenz Invariance as a local invariance we have the Minkowski (flat) spacetime metric, which then generalizes to the Riemannian metric (in General Relativity).

As with the point particle dynamics, for the field dynamics we have three ways to formulate the behavior: (1) differential equation; (2) function variation (on Lagrangian); and (3) functional variation (on the Action). We will see similar limit phenomena as before, but also new phenomena, including (i) inevitable BH singularity formation (the Penrose singularity theorem); (ii) FRW Universe formation (from homogeneity and isotropy); (iii) the BH collapse singularity; (iv) the atomic collapse radiative 'singularity'.

Classical dynamics, thus, has two field-like formulations to describe the world: field and manifold. Such formulations can be interrelated mathematically, so what is happening is more a matter of physics emphasis and convenience. The emphasis on this difference, that appears to be no difference (mathematically), is that different physical phenomenologies are at play. Field descriptions appear to work for 'matter', where the fundamental elements are spinorial. Manifold descriptions appear to work best for geometrodynamics (GR), where the fundamental elements are vectorial (or tensorial, such as the metric). Matter fields are renormalizable, thus quantizable in the standard QFT formulation (to be described in Book 5 [43]), while gravitational manifolds are not renormalizable, and have constraints (weak energy condition and positive energy condition given the existence of spinor fields on the manifold).

The presentation in Books 1-3 [40,41,46], on 'classical' physics, is partly done to make the transition to quantum physics simple, obvious, and in some cases, trivial. Consider the functional variation (Action) formulation of the behavior (whether point-particle or field), this can be captured in integral form, as was done by D'Alembert very early [7] (then by Laplace [6]). Note the use of a large constant to effect a 'highly damped' integral for selection purposes (on variational extremum of the action). To transition to the quantum theory we also have the large constant from 1/h, and so the only difference is the introduction of a factor of 'i', to effect a 'highly oscillatory' integral for selection purposes.

After the transition to a quantum theory, for the point-particle descriptions, the classical collapse problem for atomic nuclei is eliminated. The spectral predictions have excellent agreement with theory, but there is still fine-structure in the spectra not fully explained. The theory is not relativistic and some initial corrections for this are possible (without going to a field-theory) and these indicate closer agreement and explain most of the fine-structure constant discrepancy (and reveal alpha in another place in the theory). It is shown in Book 3 [41] and Book 4 [42], that the GR singularity problem, however, remains unresolved (for the test case of spherical dust shell collapse, done in a full GR analysis, then quantized in a full self-adjoint quantization analysis [42]).

In Book 5 [43], the transition to quantum theory is continued to the field theory descriptions. A precise description/agreement of atomic nuclei is now possible with QED, and within the nuclei themselves (quark confinement) with QCD. The field theories have a small set of bothersome infinities, however, which is eventually solved by renormalization [43]. As mentioned, the quantization of manifold theories, such as GR, does not appear to be possible due to non-renormalizability. Not to be deterred, in Book 6 [44] we consider a Hamiltonian description of a GR system whose quantization would involve an energy spectrum based on that Hamiltonian, if we then use analytic continuation to take us to the thermal ensemble theory based on the partition function that results, we can consider the thermal quantum gravity (TQG) of such systems.

This last example (from Book 6), showing a consistent TQG theory if we use analyticity, is part of a long sequence of successful maneuvers involving analytic continuations in different settings. What is indicated is the presence of an actual complex structure to the stated theory. There is the trivial complex structure extension mentioned above that brought us from the standard classical physics theory to the standard path integral quantum theory. But we also see actual complex structure at the component level with time complexation (that ties to thermal version of the theory by defining the partition function), and we have complex structure as the dimension-level in the form of the successfully applied dimensional regularization procedure used in the renormalization program.

As well as covering the breadth of core physics topics at both undergraduate and graduate level (for courses taken at Caltech and Oxford), including extensive presentation of problems and their solutions, the Series also examines, in specific cases, the boundaries of the physical world "from the inside" (and then later "from the outside"). To this end exploration of spherical dust collapse to form a singularity is examined in a fully general relativistic formalism, and then carried-over to a quantum minisuperspace (quantum gravity) analysis (in Books 3 and 4 [41,42]). Also examined in-depth are the topics of black hole thermodynamics and quantum field theory with alternate vacua (part of Books 5 and 6 [43,44]). The in-depth material comprises the topics covered in my PhD dissertation [81], portions of which are published [82-85].

In recent work on machine learning, that includes statistical learning on neuromanifolds [24], we find a possible new source for a foundational element for statistical mechanics (entropy) via seeking a minimal learning process/path on a neuromanifold [24]. By the time the Series reaches thermodynamics in Book 6, therefore, the foundational thermodynamics elements have all been established from the physical descriptions discovered in Books 1-5, they just haven't been put together in a comprehensive analysis that gives us the fundamental constructs of thermodynamics and statistical mechanics. That said, it would seem that thermodynamics is, thus, entirely derivative from other, truly fundamental theories. Not so, in the joining of the parts to make thermodynamics we have something greater than the sum of the parts. In the 'system' descriptions we find that emergent phenomena exist. This, at least, is unique to thermodynamics, so it is fundamental in this "sum greater than the parts' aspect.

In Book 7 (the last) of the Series, we consider the standard physical world, described by modern physics, "from the outside." In doing this we've already eliminated part of the mystery of entropy by the geometric 'neuromanifold' description. If we can understand other oddities of the standard theory, and arrive at them naturally, then we might have an even deeper dive into modern physics, testing the limits of what is possible, and see possible future developments and unifications of the theory. This is what is described in papers [70,87-90], and organized along with current results into the final Book of the series.

Efforts in the last book of the Series involve choices and concepts identified in the prior six books of the Series, and theoretical maneuvers gleaned from the most advanced courses in physics and mathematical physics taken while at Caltech (as an undergraduate and then as a graduate) and the Oxford Mathematics Institute (as a graduate), and the University of Wisconsin at Milwaukee (as a graduate).

The broad range of topics covered in the Series is, initially, similar to the Landau & Lifshitz graduate textbook series (see [27]), with a similar exposition on classical mechanics at the start of Book 1. Even with well-established classical mechanics, however, there are significant, modern, updates, such as (modern) chaos theory. In the final two books of the Series (Books 6 and 7 [44,45]) we arrive at statistical mechanics and thermodynamics, together with modern topics such as black hole thermodynamics, thermal quantum gravity, and emanator theory.

Key constants and structures of physics, their discovery from the experimental data, and their theoretical placement in the "Grand Scheme," are emphasized throughout the Series. The constant alpha, a.k.a. the fine structure constant, appears in numerous settings so special note of the occurrence of alpha will be made in each chapter. This is the case even at the outset with Book 1, due to fundamental numerical constants appearing from chaos theory. In Book 7 we see the origin of alpha, as a maximal perturbation amount, appears naturally in a formalism for maximal information 'emanation'. But maximal perturbation in what space and in what manner? In Book 7 of the series [45] we will see a possible representation of such an information entity, and its space of existence, in terms of chiral trigintaduonions.

Thus, in the end, this is an effort to tell of a journey to a special place "where many paths and errands meet", giving rise to emanator theory and an answer to the mystery of alpha. Part of this journey is equivalent to 'finding the arkenstone' (alpha) in the most unlikely of places, the trigintaduonion emanation mathematics underpinning the emanator formalism (e.g., Smaug's Lair, described in Book 7 [45]). Why I should have wandered into such an odd place (mathematically speaking), and why I should posit a deeper form of quantum propagation using hypercomplex trigintaduonions, here called emanation, is why there is such extensive background on standard topics. This extensive background even impacts the classical mechanics description via its modern chaos theory material (due to a possible relation between C_∞ and alpha). The critical role of emergent phenomena is only understood at the end, including for manifolds in geometry and neuromanifolds in statistical mechanics, and leads to a Book 6 that goes from very basic (initial thermodynamics) to very advanced (emergent phenomena). Much is made clear with emanator theory, including how reality is both fractal and emergent. At this point in the journey, as with Tolkien, this much I can say: "The Road goes ever on and on … And whither then? I cannot say".

The seven books in the Series are as follows:
Book 1. Classical Mechanics and Chaos
Book 2. Classical Field Theory
Book 3. Classical Manifold Theory
Book 4. Quantum Mechanics and the Path Integral Foundation
Book 5. Quantum Field Theory and the Standard Model
Book 6. Thermal & Statistical Mechanics, and Black Hole Thermodynamics
Book 7. Maximum Information Emanation and Emanator Theory

Overview of Book 1
Book 1 is a modern exposition of classical mechanics, including chaos theory, and including ties to later theoretical developments as well. The exposition consists, throughout, of the presentation of interesting problems with many solved, the others left for the reader. The problems are drawn from classical mechanics (CM) and mathematics courses taken at Caltech, Oxford, and the University of Wisconsin. The courses range from undergraduate level to advanced graduate level. The courses had a rich and sophisticated selection of textbook and reference material, as you might expect, and those reference texts are, similarly, drawn on here. Those classical mechanics texts, listed by author, include: Landau and Lifshitz [27]; Goldstein [25]; Fetter & Walecka [29]; Percival & Richards [28]; Arnold (ODE) [32]; Arnold (CM) [37]; Woodhouse [38]; and Bender & Orszag [39]. Notice how the first Arnold reference and the Bender and Orszag reference involve textbooks focused on ordinary

differential equations (ODEs). Likewise, an analysis of the excellent, and rapid, exposition by Landau and Lifshitz, reveals that it partly progresses through the material by going through ODEs of increasing complexity (corresponding to more complicated pendulum motion, for example, such as by adding a frictional force). This strong alignment with the underlying mathematics of ODEs is continued in this exposition, so much so that an appendix is provided for a quick review of ODEs from the applied mathematics perspective.

Particle dynamics, with and without forces, are described, with all arriving at descriptions with chaotic motion, with chaos described in the latter half of Book 1 [46]. Universally it is found that systems transitioning to chaotic behavior do so with a remarkable period-doubling process and this will be described both mathematically and with computer results. In the analysis of such dynamical systems we will find that periodic physical systems can be described in terms of repeated "mappings", e.g., classic dynamic mappings [91], and when described in this way the transition to chaos is made much more mathematically evident (as will be shown). The familiar Mandelbrot set is generated by such a repeated mapping, where it's "edge of chaos" is defined by the fractal boundary of the classic Mandelbrot image.

Properties of the classic Mandelbrot set will be relevant to the physics discussed in Book 1 and Book 7, including the property that the fractal boundary has a fractal dimension of 2 (the fractal dimension of the boundary can be between 1 and 2, to get equal to 2 is special). With the Mandelbrot set we also recover the well-studied constants associated with the universal Feigenbaum constants [19]. In the Mandelbrot set we can clearly see the fundamental constant for maximum perturbation that is at maximum antiphase (negative) with magnitude C_∞, where the same results hold for a family of basic formulations (for a variety of Lagrangian formulations, for example).

From the Lagrangian variational formulation of 'action' for particle motion we will eventually define the path integral functional variational formulation involving that same Lagrangian to arrive at a quantum description for the non-relativistic quantum particle motion (described in detail in Book 4 [42], and relativistic in Book 5 [43]). From the quantum description we arrive at the propagator formalism for describing dynamics (this exists in the classical formulation too, but typically is not used much in that context). Complex propagators will then be found to have ties to statistical mechanics and thermodynamics properties (Book 6 [44]). The ties to statistical mechanics are further emphasized when at the "edge of chaos" but with the orbit motion still confined. This may be associated with an ergodic regime, thus an equilibrium and martingale regime, the existence of which can then be used at the start of Book 6 [44] statistical mechanics and thermodynamics derivations with the existence of equilibria established at the outset. The existence of the familiar entropy measures are already indicated in the neuromanifold description (Book 3 [41]), thus, together with equilibria, the Book 6 thermodynamics description is able to begin with a well-established foundation that is not claimed by fiat, rather claimed as a direct result of what has already been determined in the theory/experiment described in the previous books of the Series.

Overview of Books 2 & 3
When moving from a theory of point particles to a theory of fields, there's not much discussion in the core physics books on fields in a general sense, it usually just directly jumps to the main field of relevance, Electromagnetism (EM). If advanced, it may also cover General Relativity (GR), as with [125]. In what follows we will cover these topics, but we will also cover the more basic fields in 1, 2, and 3D (including fluid dynamics), as well as 4D Lorentzian Field formulations (for Special Relativity), the Gauge Field formulation (thus Yang Mills covered in a classical context), and the GR geometric and gauge formulations. This establishes the foundation for the standard forces, and upon quantization (Books 4 and 5 in the Series), lays the foundation for the standard renormalizable forces (all but gravitation).

The gravitational coupling constant 'G' is a dimensionful coupling (not like with alpha in EM), and gravitation with manifold construct can be described as a gauge field construct, although not renormalizable. Gravitation, and associated geometry/manifolds, appears to relate to its own emergent structure, as will be discussed in Book 6. From the local Lorentzian geometry and Lorentzian field descriptions we also see the first of many examples where there is system information in the complexification of some parameter, here the time component. If the Lorentzian is shifted to complex time, this shifts it to being a Euclidean field, with formally well-defined convergence properties (as occurs in statistical mechanics). Complex time also shows deep connections between classical motion and associated Brownian motion (where random walk reveals pi). Thus, it should not be surprising that an emergent manifold may have complex structure such that there is also an emergent 'thermal' manifold, possibly the neuromanifold described in Book 3 and the related partition functions examined in Book 6. Just like locally flat space-time is a natural construct in GR, so too are optimization "learning" steps on a neuromanifold such that relative entropy is selected as a preferred measure, and from it Shannon entropy and Boltzmann's statistical entropy. Thus, the manifold construct appearing at Book 3 has far reaching impact into the foundations of the thermodynamic and statistical mechanical theory described in Book 6.

Before we even get to the manifold/geometry complexities of GR, however, we have already established much with the EM field part of the theory: (i) from 'free' EM without matter we get the speed of light c, Lorentz invariance, and from that special relativity and locally flat space-time; (ii) from EM with matter we get the dimensionless coupling constant alpha.

In going over field theories to describe matter, force fields, and radiation we first describe the classical field theories (CFTs) of fluid mechanics, EM, and General Relativity, with many examples shown. This is then carried over to the quantum field theory (QFT) description in Book 5. A review of the core mathematical constructs employed in CFT and QFT is given in the Appendix. Even as the mathematical physics approach grows in sophistication, we still obtain solutions via variational extrema. Thus, determining the evolution of the system from its variational optimum now becomes the focus of the effort. System 'propagation' from one time to a later time can be described by a propagator. Although a 'propagator' formulation is possible mathematically in classical mechanics (CM) and classical field theory (CF), which are shown, this is usually not done, in favor of simpler representations for the experimental application at hand. As we move to descriptions in the quantum realm, however, the use of the propagator formalism becomes typical, and when used in the path integral formulations we arrive at a compact formulation describing both the evolution and stationary-phase solution at once.

In Book 2 the focus is on classical field theory in a fixed geometry, the main physical example is EM. In this setting alpha appears, for example, in the description of an electron-positron pair: $F = e^2/(4\pi\varepsilon a^2)$ for electron-positron distance 'a' apart, where alpha appears as the coupling constant. Later, in quantum mechanics (QM), both modern and in the early Bohr model, we have that alpha $= [e^2/(4\pi\varepsilon)]/(c\hbar)$. The appearance of alpha in these situations is occurring in bound systems. If we examine EM interactions that are unbound, on the other hand, such as with the Lorentz Force $F = q(E \times v)$, here there arises no alpha parameter, nor with the early quantum mechanical analysis of such systems such as with Compton scattering. Thus, we see an early role for alpha, but only in bound systems, thus only in systems with (convergent) perturbative expansions in system variables.

In Book 3, classical field theory with *dynamic* geometry, i.e. GR, we don't see alpha at all. Instead we see manifold constructs and the mathematics of differential geometry (and to some extent differential topology and algebraic topology). Manifold constructs are entirely encapsulated in the math background given in Book 3 and the Appendix there. An application in the area of neuromanifolds (see [24]), shows the equivalent of a geodesic path in this setting is evolution involving minimum relative entropy steps. Similar to the description of a locally flat space-time we now have a description of 'entropy' increasing/evolving according to minimum relative entropy.

General relativity (GR) stands apart from the other force fields. All the other force fields are part of an adjoint representation of the standard model vis-à-vis the stability subgroup U(1)xSU(2)$_L$xSU(3). The form of which is derivable from the chiral T one-sided products described in Book 7. The standard model is uniquely obtained in this process, and with no mention of GR. Keep in mind, however, that the adjoint representation has operation on some space (hyperspinorial in case of simple octonion right-products, for example). The 'force' due to gravity is that due to manifold curvature, where the manifold construct is possibly emergent on the space of operation. Thus, the origin of the GR force is entirely different, and it will not allow quantization like the other forces, nor will its singular solutions be resolvable via quantum physics alone, as with EM in Books 4&5, but will also need thermal physics (as will be described in Book 6).

The existence of singular GR solutions, outside of specially symmetric cases (the classic Black hole solutions), wasn't firmly established until the Penrose singularity theorem [93] (awarded Nobel prize in Physics for this in 2020). Some of this material is covered in Book 3 to show how the mathematical formalism shifts to differential topology methods to describe the singularities, with examples referencing the Hawking and Ellis classic [94] and using Penrose diagrams. This, in turn, will come in handy when describing the classic FRW cosmologies with radiation and matter dominated phases (using notes from Peebles [95], Peebles won the Nobel in Physics in 2019).

The GR development would be remiss if it didn't briefly delve into cosmological models, the classic FRW cosmologies in particular. With the GR tools developed, cosmological results are examined, starting with the entry of the cosmological constant into the formalism (a candidate for Dark energy). Various observational data on galaxy rotations and universe simulations of galaxy cluster formation both indicate the existence of Dark matter. This, then, means we have new matter, non-interacting except gravitationally, and this is actually consistent with the latest observational data on the muon g-2 value [96], where the discrepancy between theory and experiment has grown to 4.2 standard deviations, where an extension in the Standard Model appears to be in the works. This is convenient as Emanator theory (Book 7 [45]), predicts such an extension.

We can thus arrive at field equations for EM, GR, and Yang-Mills Gauge Fields (Strong and weak). We can obtain wave and vortex phenomena (as hinted in fluid dynamics). We show the classical instability for atomic matter (classical EM instability) and classical gravitational instability (leading to black hole formation with singularity). From Lagrangian formulations we can then arrive at a QFT formulation (Book 5). The QFT formulation completes the QM (Book 4) cure of "non-relativistic atomic instability" with the cure of the fully relativistic atomic description of the radiative-collapse instability. Introduction of QFT also leads to new instability or infinities, but these can be eliminated by renormalization for the EM and electroweak formulations, and the Yang-Mills strong formulation, but not the GR (gauge) formulation. The current theoretical formulation in modern physics has one glaring gap, therefore: a quantum theory of gravitation. Perhaps this is not a missing element, however, if geometry/GR is a derivative phenomenon, like the field of statistical mechanics and thermodynamics appeared as derivative phenomenon when the complexified quantum propagator gives rise to a real (quantum) partition function. The hint of a deeper emanator theory suggests emergent structures of geometry and thermodynamics are arrived at in the process of emanation, with the information emanated being that of the renormalizable quantum matter fields. In Book 7 [45] a precise mathematical meaning will be found for describing maximal information emanation.

Overview of Book 4

By 1834, with Hamilton's Principle, there was a strong foundation for what is now called classical mechanics. By 1905, with Einstein's publication on the photoelectric effect [97], the rules of classical mechanics were being superseded by the new rules of quantum mechanics. The earliest appearance of quantum mechanics, however, began with the various observations of quantization of light, starting with the strange occurrence of spectral lines for hydrogen. The hydrogen spectrum was made even stranger by a precise fit to a succinct empirical formula by Balmer in 1885 [98]. This is the beginning of an amazing period of discovery. The developments of QM from introductory to advanced roughly follows that history.

The early phase of discovery for quantum mechanics moved into the modern quantum mechanics formalism with the discovery of Heisenberg of the successful application of matrix mechanics and the resultant uncertainty principle (1925) [16]. In 1926, Schrodinger showed that the problem of finding a diagonal Hamiltonian matrix in the Heisenberg's mechanics is equivalent to finding wavefunction solutions to his wave equation [17]. An interpretation of the wavefunction was then clarified in 1927 by Born [107]. Dirac developed a manifestly relativistic formalism for the wavefunction and wave-equation for fermionic matter (1928) [108]. An axiomatic reformulation of quantum mechanics was then given by Dirac (1930) [18], laying the foundation for much of modern quantum notation and for critical issues such as self-adjointness. Dirac then described a formulation of a quantum propagation path, with quantum propagator having the familiar phase factor involving the action, in his paper "The Lagrangian in Quantum Mechanics" in 1933 [109]. In essence, Dirac had obtained a single path, in what would eventually be generalized by Feynman to all paths with the invention of the path integral formalism (1942 & 1948) [110,111]. The equivalence of a quantum mechanical formulation in terms of path integrals and the Schrodinger formalism was shown by Feynman in 1948 [111].

In a path integral description, the quantum mixture state, semiclassical physics, and classical trajectories are all given by the stationary phase dominated component. A stationary phase solution that is dominated by a single path is typical for a classical system. Thus, variational methods are fundamental to analysis of physical systems, whether it be in the form of

Lagrangian and Hamiltonian analysis, or in various equivalent integral formulations.

Feynman's discovery of the path integral formalism wasn't solely based on the prior work of Dirac (1933) [109], although by appending that paper to his PhD thesis (1946) its importance was clearly emphasized. Feynman also benefited from work going as far back as Laplace [6] for selection process based on highly oscillatory integral constructions that self-select for their stationary phase component. This branch of mathematics eventually became associated with Laplace's method of steepest descents, then to the work of Stokes and Lord Kelvin, then to the work of Erdelyi (1953) [112-114].

Feynman and others then invented quantum field theory for electromagnetism (QED) during 1946-1949 (more on this later). Extension to electroweak occurred in 1959, and to QCD in 1973, and to the "Standard Model" in 1973-1975. Thus, the impact of the path integral revolution in quantum physics was felt well into the 1970's, but this was only the beginning. At their inception path integrals were examined by Norbert Wiener, with the introduction of the Wiener Integral, for solving problems in statistical mechanics in diffusion and Brownian motion. In the 1970's this led to what is now known as "the grand synthesis" which unified quantum field theory (QFT) and statistical field theory (SFT) of a fluctuating field near a second-order phase transition, and where use of renormalization group methods enabled significant advances from QFT to be carried over to SFT.

The grand synthesis is one of many instances to come where we see analytic continuation of a constant or a parameter giving rise to familiar physics in the thermodynamic and statistical mechanics domains, showing a deeper connection (still not fully understood, see Book 7). The Schrödinger equation, for example, can be seen to be a diffusion equation with an imaginary diffusion constant. Likewise, the path integral can be seen to be an analytic continuation of the method for summing up all possible random walks.

In Book 4 we also carefully examine the closest gravitational equivalent to the hydrogenic atom (dust shell collapse). What results is an incomplete formulation due to boundary conditions, where to get the time choice you must input that time choice. No specific choice of time is indicated to avoid infall-collapse. The results, however, can show stability and consistency in a "full" thermal quantum gravity description where analyticity is employed. Success in this way, and not others, suggests possible fundamental role of analyticity and thermality (Books 6&7) and also suggests that thermal quantum gravity TQG may 'exist' or be well-formulate-able, while quantum gravity QG generally might not 'exist'. These results, shown in Book 6, provide the lead-in to the Book 7 discussion on Emanator theory, where core concepts in Books 1-6 that tie to emanator theory are brought together in a new theoretical synthesis.

Overview of Book 5
In Book 5 we show QFT's in the gauge field representation, which clearly relates the choice of field theory to a choice of Lie algebra, which, in turn, can be related to a choice of group theory (such as U(1) and SU(3)). From this we can see that non-classical algebraic constructs are ubiquitous in QM and QFT, so a review of Group Theory and Lie Algebras is given in the Appendix, as well as a review of Grassman Algebras, and other special algebras needed in QM and QFT. Similarly, as regards choice of approach, we find that the Schrodinger and Heisenberg formulations often provide the only tractable way to get a solution for bound systems. In critical theoretical considerations, however, the path integral approach is best (as will be shown). In seeking a deeper theory, the more unified path integral (PI) approach provides important hints as to a deeper theory (see Book 7).

In Book 5 we get the highest precision result for the value of alpha, in its role as perturbation parameter. If a calculation of the electron magnetic moment parameter g-2 is performed, with all of the Feynman diagrams appropriate to expansions up to 5th order, we get a determination of alpha up to 14 digits, where 1/alpha=137.05999...... . This gives us one of the most precise measurements of alpha known. When a similar analysis is done for the muon g-2, given the much larger muon mass, particle production pairs of other particles have a measurable effect, and we are able to probe the lower masses of the standard model that are present. In doing this, in preliminary experiments, there is a discrepancy indicating more particles, e.g. the Standard Model will need to be extended (possibly with a type of 'sterile' neutrino). These missing particles could be the missing "Dark Matter". The prediction of such in Emanator Theory, and why there should be an imbalance between the left and right neutrinos (hint: maximum information transmission) is described in Book 7.

Part of the description of quantum field theory entails use of analyticity and other complex structures to encapsulate more of the physics in a complex extension to the space (or dimension). This often leads to formulations in terms of complex integration, with the choice of complex contour specified, such as with the Feynman propagator. One of the main renormalization methods, for example, is to use dimensional regularization, which entails analytically continuing expressions with dimensionality to dimensionality as a complex parameter. There is also the aforementioned shift to complex and to "Wick rotate" expressions with real time to expressions with pure complex time. In doing this the statistical mechanical partition function for the system is obtained, with well-defined summation. Thus, a connection between 'thermality' and complex structure, in the time dimension at least, is indicated.

The second part of Book 5 describes QFT on curved space-time (CST), where we arrive at an early analysis of Black Hole thermodynamics. Here we find that space-time curvature gives rise to thermality and particle production effects. Black Hole thermality was revealed in Hawking radiation [118], due to the causal boundary at the horizon. Such thermality is even seen in flat space-time (Book 5) if causal boundaries are induced, such as in the case of an accelerated observer [143].

QFT on CST has one further gift, critical to the statistical mechanics formalism to follow in Book 6, and that's the spin-statistics relation. This relation is usually assumed, along with other critical notions, such as entropy, and the relation between entropy and density of states. These are all shown, with the presentation path chosen in this Physics Series, to be fundamental or derivative to the formalism already established in Books 1-5 (to prepare for Book 6).

The choice of time is related to choice of vacuum, which is related to choice of field geometry or observer motion (such as constant acceleration or expansion). If you have flat spacetime QFT with a boundary, then you have thermodynamic effects (e.g., the Rindler observer). In this setting we can compare the Hawking derivation of Hawking Radiation using the Euclideanization 'trick' vs the Bogoliubov transformations of the field to the Rindler geometry from the Minkowski geometry (if chosen as the asymptotic vacuum reference). With QFT on CST we also arrive at spin-statistics as mentioned, and get the final extension of the theory by way of Grassman algebras, to arrive at thermodynamically consistent Bose and Fermi statistical descriptions on quantum matter.

Overview of Book 6

Thermodynamics is the oldest of the physics disciplines (fire), with unapologetic use of phenomenological arguments and mysterious thermodynamic potentials (entropy). Obviously, thermodynamics is still prevalent today, including in its more quantified form via statistical mechanics. How is this not a failure of the mechanistic description of the universe indicated by CM and even QM? Concepts that appeared in QM, such as probability, are now occurring again. Other new concepts appear as well, including: approximate statistical laws; equations of state; heat as a form of energy; entropy as a variable of state; existence of equilibria; ensembles/distributions; and existence of the partition function. Many of these concepts appear in the path integral descriptions with the analyticity methods/extensions mentioned previously, so there are hints of a deeper theory that arrives at much of thermodynamics/Statistical mechanics foundation from the existing quantum theory.

Book 6 has been placed after the other chapters to await identification of entropy as fundamental in that it can be identified as an intrinsic system function even before getting to thermodynamics. We also already have experience with many particle systems, via QFT (especially in CST where particle creation is almost unavoidable), without directly tackling that scenario (due to QFT effectively already being many-particle, with analytic determination of many-particle system functions, such as entropy). With entropy presented at the outset as an important system variable, the derivation of thermodynamic potentials is then a straightforward process, as will be shown. The standard SM connections to thermodynamics can then be given. Thus, in covering Thermodynamics and Statistical Mechanics we start with the foundations of the theory mostly established, such as entropy (also with equipartition equivalent to sum on paths with no weightings, etc.), with no assumptions. Everything follows directly from the theoretical discoveries outlined in the preceding books in the Series. We don't see new connections to alpha, but we do see new structures/effects, especially manifold constructs (as with GR, where we also saw no role for alpha).

The close ties between QM Complexified giving rise to a particle ensemble partition function, and QFT complexified and field ensemble partition function, is now simply a derivative aspect of the fundamental complexation posited. This complexation will be posed in Book 7 with emanation in a complexified perturbation space.

From Atomic Physics, described in Book 4, we also obtain the standard rules on electron shell completion (that is encoded in the periodic table). Similarly, we can also understand the origins of the intermolecular quantum chemistry rules. When taken to the statistical mechanics (SM) extreme we have thermodynamic equilibrium emergent from (the Law of Large Numbers (LLN) and reverse Martingale convergence. With completion of application to chemical processes we have clear phase-transition effects, as well as equilibrium and near-equilibrium effects. The familiar chemistry results, with phases of matter.

From chemical equilibrium and near-equilibrium, with 10^{23} elements that interact weakly or not at all, we have two generalizations. The first is to consider chemical near-equilibrium and directly obtain an emergent process at this level, this is the branch that gives us biology/life at its most primitive level. The second is to consider equilibrium and near-equilibrium in general when the elements interact strongly (with 10^{10} elements, say), this is the branch that describes biology/life at its most advanced social level and economics. In classic shot noise, the granularity of low-current flow (due to discreteness off electron charge) leads to a noise effect. Thus, as we consider situations with fewer elements, there are more complications, not less, due to granularity noise effects, and we enter the realm of machine learning with sparse data. Noise effects can be significant in complex systems, especially in biology where it is part of what is selected (such as in hearing, for background noise cancellation).

The second part of Book 6 explores the role of thermodynamics in efforts to extend to TQFT and TQG. This is done by exploring Black Hole settings. The recognition of a role for complex structure on system variables becomes apparent in this process (on top of the generalization to non-trivial algebras as already revealed).

In Book 6, part 2, we examine the Hamiltonian thermodynamics of some black hole geometries with stabilizing boundary conditions. In this foray into directly exploring a thermal quantum gravity (TQG) solution we assume a path integral form for the GR problem and shift directly to a partition function (by 'Wick rotation' mentioned above). We see that TQG is possible, where positive heat capacity shows stability. Another encouraging result as to an eventual unifying theory comes from String theory via its explanation of BH thermodynamics and BH horizon effects with the BH fuzz solution (via use of the holographic hypothesis and the related AdS-CFT relation [120,121]).

In Book 6, part 2, we also examine the propagator to partition-function transformation upon complexation, which leads to a thermodynamic theory for some equilibrium formulation, with certain parameter settings required for stability (positive heat capacity). This is doable in a variety of settings, suggesting how such thermodynamically consistent boundary conditions may be what constrains the classical motion and BH singularity formulation by the effect of this stabilization manifesting for certain internal geometries. Successful TQG (Thermal Quantum Gravity) formulations, such as for RNadS and Lovelock spacetimes shown in Book 6, via reformulation using analyticity, and not via non-analytic approaches, suggests a possible fundamental role of analyticity once again and also suggest that TQG may 'exist' or be well-formulate-able, while QG generally might not 'exist'. These results, together with core concepts from Books 1-6 that tie to emanator theory, are brought together in a new theoretical synthesis in Book 7.

Overview of Book 7
In Books 4,5, and 6 of the Series, we explored examples of QM with imaginary time, QFT in CST, Thermal QFT, minisuperspace QG, and Thermal QG. In this effort we find the path integral, and PI propagator, to provide the most general representation. In seeking a deeper theory in Book 7 we build on the sum-on-paths with propagator formulation to arrive at a sum-on-emanations with emanator formulation.

Propagation in a complex Hilbert space, in a standard QM or QFT formulation, requires the propagator function to be a complex number (not real or quaternionic, etc., [122]). This prohibits what would otherwise be an obvious generalization to hypercomplex algebras. In order to achieve this generalization, we have to introduce a new layer to the theory, one with universal emanation involving hypercomplex algebras (trigintaduonions) that is hypothesized to project to the familiar complex Hilbert space propagation with associated fixed elements (e.g., the emanator formalism projects out the observed constants and group structure of the standard model). The 'projection' is an induced mathematical construct, like having SU(3) on products of octonions, but here it we be the standard model U(1)xSU(2)xSU(3) on products of emanator trigintaduonions. Thus, in Book 7 a unified variational formulation is posed, one that arrives at alpha as a natural structural element, among other things, uniquely specified by the condition of maximal information emanation.

In Book 7 we also make note of the implications of a fundamental mathematical operation on a space that is repeated or added. The non-GR forces are given by the form of the operation (the sequence forming an associative algebra), the GR forces are given indirectly by the form of the space, this leaves the aspect "repeated or added" to be considered with care. If a purely 'repeated' operation, or mapping, occurs we can return to the dynamical mapping discussion of Book 1, where chaos can occur and is ubiquitous. There, the primal 'phase transition', the transition to chaos, is evident. If an operation with addition is involved (in the statistical sense of multiple elements), along with repeated overall steps, we arrive at the general framework of statistical mechanics with effects from the Law of Large Numbers (LLN) and reverse Martingale convergence, among other things (Book 6). Most notable, however, is the prevalence of a new effect, that of phase transitions and the emergence of new structure (order from disorder), including the remarkable structures of chemistry and biology.

Why the recurring 'Cabbalistic formula'? was a question even in the time of Sommerfeld [58]. Now, the numerological parallel is more exact than realized at that time, so is too much a coincidence to be by chance. The non-coincidence appears to be due to the maximal nature of information transmission in a variety of circumstances (in physics, biology, and even human communication with sufficient optimization) as well as with the fractal-like repetition of key parameter sets that occurs in these different settings $\{10,22,78,137 \cong 1/alpha\}$. We see that 10 expresses the dimensionality of propagation (or nodes of connectivity), while 22 corresponds to the number of fixed parameters in the propagation (in Book 7 we explore propagation in a 10 dimensional subspace of the 32 dimensional trigintaduonion space, leaving 22 dimensions at fixed values that appear as parameters in the theory). We will see the number 78 relates to generators of the motion, and that there are 4 chiralities of motion ('doubly chiral'). We will also see that 137 is simply the number of independent tri-octonionic product terms in the general chiral trigintaduonion 'emanation'.

Synopsis – Frodo Lives

Tolkien wrote of eucatastrophes [127], perhaps he anticipated the constructive role of emergent phenomena in maximum information transmission.

Preface to Physics Series, Book #1, on:

Classical Mechanics and Chaos

This book provides a description of classical mechanics, starting with the classic formulations of point particle motion. The first approach to doing this was using differential equations (Newton's 1st and 2nd Law); the second was using a variational function formulation to select the differential equations (Lagrangian variation); the third was using a variational functional formulation (Action formulation) to select the variational function formulation. This book will describe the three formulations and solve problems in each.

It wasn't until classical mechanics was already well established that it was realized that there are two domains for motion in many systems: non-chaotic; and chaotic. This is a modern exposition of classical mechanics, thus including chaos theory, and including ties to later theoretical developments as well. The exposition consists, throughout, of the presentation of interesting problems with many solved, the others left for the reader. The problems are drawn from classical mechanics and mathematics courses taken at Caltech, Oxford, and the University of Wisconsin. The courses range from undergraduate level to advanced graduate level. The courses had a rich and sophisticated selection of textbook and reference material, as you might expect, and those reference texts are, similarly, drawn on here. As we progress through the material we will see that we are effectively studying ordinary differential equations (ODEs) of increasing complexity (corresponding to more complicated pendulum motion, for example, such as by adding a frictional force). This strong alignment with the underlying mathematics of ODEs motivates the placement of an appendix for a quick review of ODEs from the applied mathematics perspective.

In addition to a modern exposition of the underlying ODE theory, with chaos included, the other main modern elements are to indicate where the classical mechanics theory can bridge into the theories yet to come, such as quantum mechanics and Special Relativity. There are five theoretical implementation areas of Classical Mechanics where Quantum Mechanics is trivially indicated (by analytic extension/continuation, or by algebraic modification from abelian to non-abelian), and such areas are described in detail. Similarly, there are three areas of experimental application where Special Relativity is indicated, that are also described.

Chapter 1. Introduction

This book provides a description of classical mechanics, starting with the classic formulations of point particle motion. The first approach to doing this was using differential equations (Newton's 1st and 2nd Law); the second was using a variational function formulation to select the differential equations (Lagrangian variation); the third was using a variational functional formulation (Action formulation) to select the variational function formulation. This book will describe the three formulations and solve problems in each.

In a description of particle motion, assuming not in a parameter domain with chaotic motion, several important limits are found to exist. Examples include: the universal constants from the aforementioned chaos phenomenon, that are still encountered in non-chaos regimes if driven "to the edge of chaos". Scatting is defined in the asymptotic limit and perturbation theory is well-defined in the sense that it is convergent. Overall, if the evolution is described as a 'process' it is often a Martingale process, which has well-defined limits. So, we have descriptions for motion, typically reducible to an ODE, and for which solutions (requiring limit-definitions) are typically found to exist.

The development of classical mechanics mostly occurred during the years spanning 1687 to 1834 [1-13]. There was then a sizable gap while other discoveries were made, ranging from quaternions [14,15] to electromagnetism, to quantum mechanics [16-18]. Finally, in 1976 the last key element of the classical theory was revealed with the discovery of chaos universality [19]. Also, during this time, more sophisticated mathematical approaches became more common [20,21].

A major theory departure from classical mechanics occurred with special relativity, which was revealed by the discovery of the Lorentz Transform in 1899 [X] (there were early hints in the studies of Fizeau [22] in 1851, but this was not understood until Einstein decades later [23]). Development of classical mechanics methods is still very relevant in present day, partly due to related developments in modern AI. One of the strongest classification methods known, the Support Vector Machine (SVM), for example, is based on a classical mechanics (Lagrangian)

1

formulation in a control theory application (with inequality constraints) [24].

A modern textbook description of classical mechanics without chaos theory can be found in Goldstein [25]. A key development in the theory, in terms of variational invariants, was contributed by Noether in 1918 [26]. Other modern textbooks drawn upon in this book include the classics by Landau and Lifshitz [27], Percival & Richards [28], and Fetter & Walecka [29]. Two-timing analysis [30] and stability analysis [31,32] is also included in this work followed by the aforementioned critical developments in chaos theory [19,33,34] and the critical appearance of fractals [35,36]

This is a modern exposition of classical mechanics that consists, throughout, of the presentation of solutions to interesting problems from a number of classical mechanics texts, including: Landau and Lifshitz [27]; Goldstein [25]; Fetter & Walecka [29]; Percival & Richards [28]; Arnold (ODE) [32]; Arnold (CM) [37]; Woodhouse [38]; and Bender & Orszag [39]. Notice how the first Arnold reference and the Bender and Orszag reference involve textbooks focused on ordinary differential equations (ODEs). Likewise, an analysis of the excellent, and rapid, exposition by Landau and Lifshitz, reveals that it partly progresses through the material by going through ODEs of increasing complexity. This strong alignment with the underlying mathematics of ODEs is continued in this exposition, so much (so that an appendix is provided for a quick review of ODEs from the applied mathematics perspective).

Starting with Newton's differential equation F=ma, we progressively encounter more complex differential equations. Reducing a dynamical system to a set of differential equations is no simple matter, and learning Lagrangian analysis to do this will be the focus initially, but the end result can always be taken to be a form in terms of an ordinary differential equation (ODE), or set of such. So we can reduce the problem of describing the motion of a system to that of solving an ODE, does that mean we are done? For simpler ODEs, yes, analytically in fact (in the Appendix we see, for example, that second order linear differential equations with constant coefficients can always be solved). For more complex ODEs, still yes, but computational tools are needed (solution not in closed form). Sometimes ODE's demonstrate instabilities, however, and for these more sophisticated analysis is needed and there may not be simple answers (such as the existence of the strange attractor phenomenon) [37]. More revolutionary than mere instability is the discovery of chaos. An ODE might be well-behaved in one regime but may shift to 'chaotic motion' in another regime. The "edge of chaos" is marked by a universal period doubling behavior and is described in Ch. 7. Everything an ODE specialist might have feared could occur, insofar as complexity is concerned, is found to be the case (with instabilities and strange attractors, etc.), and then this was doubled with the discovery of the new phenomenon of Chaos via Universality. For the ODE examples described here, the focus is on physics problems, so the chaotic solutions directly relate to chaotic motion.

In addition to a modern exposition of the underlying ODE theory, with chaos included, the other main modern elements are to indicate where the CM theory can bridge into the theories yet to come, such as quantum mechanics (QM) [42] and Special Relativity (SR) [40]. For perturbation theory involving solutions to an ODE, a variety of techniques are shown. If complex analysis is used, we get solutions, for example, but we also glimpse the general ODE problems encountered in QM. The general ODE's described in the Appendix arrive at the Sturm-Liouville form, for example, which has a self-adjoint formulation relevant to QM. Even more general is the Navier-Stokes equation (relevant to fluid dynamics), and more general than that is the N-S equation without species conservation (like in a semiconductor where there can be carrier generation, thus no conservation, with a modified continuity equation, etc.). The couplings required in the relativistic formulation, in turn, create quite a complicated mess that is almost never solved directly without approximation. In practice, the 'master Navier-Stokes equation' is approximated within some realm of operation that is relevant.

In what follows, there are five theoretical implementation areas of Classical Mechanics (CM), where Quantum Mechanics is trivially indicated (by analytic extension/continuation), and such areas are described in detail. Similarly, there are three areas of experimental application, where Special Relativity is indicated, and those are also described.

1.1 The *sine qua non* of chaos and emergent phenomena
It will be seen that classical mechanics is a special case of a larger quantum mechanical theory, thus it might seem that we've demoted classical mechanics to be a theory that is derivative of another... *but for* the existence of chaos theory. Chaos is a fundamentally new dynamical

3

aspect (of all theories classical, quantum, statistical, with appropriate differential form), but it is its simplest (while still being familiar) in the classical mechanics regime. Chaotic motion is exhibited ubiquitously, but can also be avoided in many classical mechanics problems, such as small oscillation problems. Chaos, as a universal phenomenon, also has universal constants, which will be explored. A simple pathway to finding chaos is to use the Hamiltonian representation and examine any periodic motion involving nonlinearities. When viewed as an iterative map, chaos domains are then clearly exhibited (as will be shown in Ch. 7). Similarly, statistical mechanics might be viewed as a derivative theory of classical mechanics, *but for* the occurrence of the entropic measure and of emergent (phase transition) phenomena (to be discussed in other books in this Series [40-46], especially [41] and [44]).

1.2 The role of ordinary differential equations (ODE's), phenomenology, and dimensional analysis

A perusal of the table of contents will reveal many sub-sections relating to application of ODE's. This focus on ODE's is not by accident and neither is the inclusion of a large appendix (App. A) on ODE's. (App. A will describe general ODE methods, and advanced methods, with numerous worked-out solutions.) Almost always, the classical mechanics problem can be reduced to solving an ODE. Since this is what we started with, with Newton (a 2nd order ODE), this might not seem like progress, however, arriving at the correct ODE for a system is often difficult if not well-nigh impossible without the intervening techniques (Lagrangian and Hamiltonian). So, such methods are obviously needed, it's just that a deep knowledge of ODE's is needed as well. Knowing that we will have a differential equation, and restricting to equations consistent with dimensional analysis, we can often directly arrive at the basis for a number of phenomenological arguments for equations of motion and their solutions via ODE's (and suggestions or explanations as to new phenomena). Dimensional analysis and phenomenology are described in Ch. 9.

1.3 Sources of problems; Level of coverage; Detailed solutions; Advanced Methods

Some of the problems (with and without solutions) are at the level of PhD candidacy exam questions (an exam, or "preliminary exam," that is taken at end of the second year of a Physics PhD program in order to advance to candidacy, at some institutions, such as UWM and U. Chicago). Such problems tend to be the most difficult. Some of the problems, almost as difficult, are related to problems I was assigned in undergraduate and graduate course taken while a student at Caltech. In many cases my carefully worked-out solutions were used in the "solutions sets" provided to the class later. Such problems and my solutions are shown for problems from the following Caltech (ca 1987) courses: Topics in Classical Physics; Advanced Dynamics; and Methods of Applied Mathematics (in App. A). Often the problems, or examples, in the coursework were derived from problems from the main textbooks available in Classical Mechanics. Thus, such sources were also directly drawn upon for some of the problems solved here as well, and include solutions for problems from the following classic texts: Goldstein [25]; Landau&Lifschitz [27]; Percival&Richards [28]; and Fetter&Walecka [29]. Solutions are provided in extensive mathematical detail, like what might be provided in a class lecture, in order to teach solution technique (index "gymnastics") in detail.

1.4 Synopsis of Chapters to follow
To begin, we consider the classical theory of point particle motion and classical mechanics. This starts, in Sec. 2.1, with a brief description of Newton's calculus formulation (1687) [1], where the Newtonian force equals mass times acceleration (a second derivative on position in the notation of Leibnitz). Leibnitz was the other major inventor of calculus, with use of integral calculus in unpublished notes in 1675 [2], and published in 1684 (for translation see Struik [3]). Leibnitz also described the fundamental theorem of (modern) calculus (the inverse relation between integration and differentiation) in 1693 [4]. The early role of math-oriented polymaths in the development of the mathematical foundations of classical mechanics continued with Euler and Laplace. Euler made contributions early, with Mechanica (1736) [5], but continued with developments in the underlying mathematics and mathematical physics for several decades, impacting Lagrange more than fifty years later, in 1788 (with the synthesis known as the Euler-Lagrange equations). Laplace's method described in (1774) [6], similarly, had a major impact on Hamilton's reformulation in 1834 (which gives rise to the classical propagator associated with $\int e^{Mf(x)} \, dx$, for $M \gg 1$) [6], as well as path integral methods in the 1940's (quantum propagator associated with $\int e^{iMf(x)} \, dx, M \gg 1$) [48].

After Newton, the next major formulation of the classical theory was with D'Alembert's description of force in the context of virtual work (1743) [7]. Virtual work, balancing to zero work actually done, is equivalent to a form of the Euler-Lagrange equations [8,9], which reacquire the equations of motion as before but now with a much easier description of holonomic constraints (such as for rigid bodies, where constraint equation is not a differential equation). In Sec. 3.3.1 we review the types of constraints, such as holonomic. In many situations we have non-holonomic constraints (such as for a rolling object). The complication of non-holonomic constraints is easily managed in Hamilton's reformulation in terms of the Principle of Least Action (1833,1834) [10-13], described in Ch. 3. Hamilton shifts the mathematical underpinning of the theoretical formulation to be a variational extremum of an action functional defined as the integral of a Lagrangian function for a point particle over time (along a trajectory or path). The variational minimum, e.g. the least action principle, then recovers the Euler-Lagrange equations to describe the same equations of motion as with D'Alembert, except now we have the means to handle non-holonomic constraints by way of Lagrange multipliers (briefly described in Sec. 3.3.1, and then used in some examples in Sec 3.3.2). Hamilton also co-discovered quaternions (1843-1850) [14], along with Olinde Rodrigues (1840) [15], which would be used in expressing early electromagnetism by Maxwell (to be discussed in [40]), and in indicating more complex algebras (a prelude to quantum mechanics – to be discussed in [42]).

The variational formulation shown in Ch. 3 also 'unifies' the classical theory in other ways [7-14], as well as bridging to the "new" quantum theory (details in [42]). This is because the quantum theory can be expressed in terms of an oscillatory integral formulation, where the constraint to have a minimal action is arrived at not as a fundamental variational rule, but as a consequence of summing over all paths of motion whose actions enter as phase terms in a highly oscillatory integral (initial math development from Laplace's method [6]), that in turn selects the classical equations of motion as a zeroth-order approximation to the oscillatory integral (stationary phase). At first order we have semi-classical effects, and a sum of the full quantum description gives the full quantum theory (see [42] for further details).

Ch. 3 specifically explores the application of the minimal action formulation in terms of a functional (the action) on the Lagrangian function integrated along a specified path. A wide range of classical systems can be described with such an application of the variational methodology. There are two main ways to formulate the action functional that are related by Legendre transformation: (i) the aforementioned

Lagrangian method and, (ii) the Hamiltonian method. The Hamiltonian, to be described (with applications) in Ch. 6, is associated with conserved quantities of the system, if they exist, such as the energy. In this latter sense, of describing the conserved quantities of the system, the Hamiltonian is introduced in Ch. 3, to express those conserved quantities in the solutions. The analysis from the perspective of a full Hamiltonian variational analysis, however, is not done until Ch. 6. The intervening very brief sections include Ch. 4 Classical Measurement; and Ch. 5 Collective Motion.

Ch.'s 3, 6, and 8, describe the first-order Hamiltonian formulation in terms of canonical coordinates. The phase space representation of the system dynamics in terms of the canonical coordinates then allows the properties of the Hamiltonian to be explored when seen as a mapping function on a phase space. We find that such mappings are area conserving and allow us to describe the asymptotic system behavior with ease in many situations, including situations that clearly demonstrate a radically new phenomenon: 'chaos'. The ubiquitous occurrence of chaos, and of classical systems "at the edge of chaos", is then described in Ch. 7.

The "universality" of chaos was shown in Feigenbaum's 1976 paper [19]. This Universality occurs with the assumption that the mapping function has a quadratic (parabolic) local maximum. Feigenbaum indicates this is a normal relation but does not elaborate further. It turns out that having a quadratic form for the local maximum (near a critical point) is a general property from the calculus of variations and Hilbert Spaces known as the Morse-Palais lemma [20,21]. The assumption underpinning the universality of chaos is valid if there exists a smooth enough function near critical points of interest, e.g., that there exists a manifold description (with a smooth function). Suppose we turn this on its head (as will be done in [47]) and suppose that chaos is a fundamental limit, always present. If this is true, then Morse-Palais must always be applicable, thus we have a manifold (geometry). This is interesting because before we even get to dynamic fields/geometries (manifolds) in [41] we see evidence of such a mathematical construct existing as a consequence of the universality of, well, Universality [19].

Ch. 8 goes into more explicit properties of canonical coordinates and transformations between them. This allows canonical coordinates to be chosen that greatly simplify the analysis by decoupling the equations of motion and making them constants of the motion, or coordinates of the

7

motion, in many cases. The most decoupled case is described by what is known as the Hamilton-Jacobi equation, which, when shifted to the operator formalism for the quantum theory, described in [42], becomes the familiar Schrödinger equation. Another formulation, in terms of appropriately chosen canonical variables, gives rise to the Poisson Bracket formulation. This is also discussed, not for its application in classical physics *per se*, but due to its trivial shift to an operator commutator formulation to arrive at the other (the first) quantum reformulation of the classical theory (the Heisenberg formulation). Ch. 9 continues with another advantage of the Hamiltonian formulation, a conserved quantity in many systems, via its application to perturbation theory. The use of Hamiltonians in both classical and quantum *perturbation* contexts are discussed. Ch. 9 also describes dimensional analysis, which when taken together with an analysis of conserved quantities, can give rise to surprising solutions based on self-similarity alone – with a few classic examples given. Extra exercises are placed in Ch. 10.

The classical mechanics described in this book only briefly touches on special relativistic corrections, i.e., it is focused on particulate matter moving at non-relativistic speeds. Thus, in this book there is the approximation of absolute time, a notion of simultaneity, and of instantaneous transmission of force with changing source position. Note that this separation of special relativity from the classical physics of this book is also reasonable, physically, in that at the level of particulate, non-relativistic, matter examined there is little opportunity to see special relativistic effects. See Sec. 3.3.2 for an early experimental indication of the existence of a 4-vector magnitude for energy-momentum in the Compton scattering formula. Another example where relativistic effects were seen, although not realized at the time, was in the experiments of Fizeau on light propagation through flowing water (1851) [22]. (Einstein remarked that "the experimental results which had influenced him most were the observations of stellar aberration and Fizeau's measurements on the speed of light in moving water" [23].) The Fizeau experiment (Sec. 4.3) gives rise to a relativistic velocity 4-vector addition calculation (for the relativistic Doppler effect). Once the relativistic Doppler effect is revealed, all of special relativity can be recovered by means of the Bondi K-calculus (described in [40]).

Once we get to notions of dynamical force fields in [40], the Lorentz transformation on Maxwell's equations (as 4-vectors) is revealed (1899), and extension of these transformations to all matter *a la* Einstein then follows in 1905. For this reason, the theory of special relativity and background and problem solutions are placed in [40] on Fields.

Thus, the fields described in this book, if at all, are static or stationary, where discussion of their general dynamical role is deferred to [40]. The classical mechanical systems considered are also simple in that only a few elements are interacting and in motion at any given time. The connections to systems with many elements is mainly left to [44] on Statistical Mechanics. Even at the classical mechanics level, however, we can still see preliminary signs of new phenomena (due to emergent Martingale phenomena, and Law-of-Large-Numbers, LLN, behavior). From this we can begin to see that there are new fundamental parameters, such as entropy (discussed in [41], in regards to information geometry, and in Book 6 [X] on Statistical Mechanics).

Note that before we arrive at [44] on Statistical Mechanics, where the fundamental role of entropy is mainly explored, we will have already 'discovered' entropy in the context of the statistical learning theory on a neuro*manifold* (given in [41]. When statistical learning is performed on a neural net (NN) construction with NN-learning via Expectation/Maximization, the learning process can be described using information geometry. Information geometry is a differential geometry formalism applied to families of distributions in statistical learning processes. In optimal statistical learning it can be shown that entropy is selected for 'local' notions of distributional distance in a similar process to Euclidean distance (flat space-time) being selected as a local geometric notion of manifold distance. In this way, entropy is singled out as a local measure just as locally flat space-time is selected (with local Minkowski metric). Aside from the theory connection, direct implementation of Statistical Learning, in the form of AI-based SVM learning [24], is actually an exercise in Lagrangian optimization with non-holonomic inequality constraints (see [24]), so will be directly accessible to those that have mastered the material in this book.

Now to begin... with Newton.

Chapter 2. Newton, Leibnitz, and D'Alembert

Mathematical descriptions of physics must attempt to justify why their description should be a certain way or evolve a certain way, amongst all of the mathematically expressible possibilities. The answer, especially in the aftermath of the philosophy espoused by Maupertus and Leibnitz [2], is typically some form of optimum selected on the state or the path of the motion (shortest path, for example). Given the idea of seeking a variational extremum, it then makes sense that there would be the invention (or discovery) of variational calculus.

Prior to 1660, pre-calculus physics had acquired a body of observational data but did not have the mathematics invented yet to contend with describing trajectories and extremal paths (which those trajectories will be shown to be). That's not to say that a body of critical mathematics development hadn't already occurred, going as far back as the invention of primitive trigonometry with the concept of the sine of the angle (sine was used in star tracking by Indian astronomers, Gupta Period, but the use of the method could trace back to the ancient Babylonians with future discoveries [75]).

Newton's fluxional calculus was invented in 1665-1666 (during the London plague), but he avoided direct use of infinitesimals in expressing his conclusions. Leibniz's calculus accepts the use and validity of infinitesimals at the outset, and began the notational development for infinitesimals in 1675 that is still in use today. Formal mathematical validity of using infinitesimals had to wait until 1963 for "Non-standard analysis' by Abraham Robinson [76,77].

The mathematical physics description of reality, thus, became established with the development of calculus in the 1660's [1,2]. Variational calculus, specifically, provides physical solutions and descriptions of reality that conform to observation, where the physical description of reality is in the form of a variational extremum [6,10,11]. This is described in detail in Classical Mechanics (CM) and Classical Field Theory (CFT). Having a variational process to select the optimum often devolves to solving some form of differential equation (reviewed in detail in the Appendix). This is fine if you can solve the differential equation, but if you can't it is beneficial to have some other analysis methodology to select equations of motion. Thus, it was recognized very early on that you could have a selection process based on highly oscillatory integral constructions that self-select for their stationary phase component [6]. This latter path will eventually lay the foundation for the Path Integral approach to quantum physics (see [42]), and to all of the classical physics that came before as a special case.

Introduction of mathematical physics concepts before formal mathematical validation is a recurring theme in physics. Another such instance is the introduction of the delta function by Dirac, formalized via L^2 distribution theory [78] (this is what is critically needed in the underlying, self-adjoint, quantum formulation).

2.1 Newton's Force Law and, with Leibnitz, Invention of Calculus

Let's begin with a restatement of Newton's three laws:

1st Law: $\frac{dp}{dt} = 0$ if $F = 0$, where $p = mv$ and m is mass, and v is velocity.

2nd Law: $\frac{dp}{dt} = F \rightarrow F = ma$.

3rd Law: The force exerted between two objects is equal and opposite.

$$\text{(Eqn.s 2-1)}$$

And, when there is more than one particle, we have for the equation of motion for the ith particle:

$$\sum_j \vec{F}_{ji} + \vec{F}_i = \dot{\vec{p}}_i ,$$

$$\text{(Eqn. 2-2)}$$

where $\vec{F}_{ji}$ is the force of the jth particle on the ith particle ($\vec{F}_{ii} = 0$), $\vec{F}_i$ is the net external force on the ith particle, and $\dot{\vec{p}}_i$ is the time derivative of the momentum of the ith particle. Recall Newton's 3rd Law, where the force exerted between two objects is equal and opposite, i.e. $\vec{F}_{ji} = -\vec{F}_{ij}$. This is referred to as the weak law of action and reaction [25].

In Ch. 1 Problem 6 (pg 31) of Goldstein [25], outlined below, we find that the standard equations of motion for the center-of-mass position and momentum, taken as starting point, not only indicates the weak law of action and reaction, but also the strong law, *where the forces strictly lie along the line joining the objects.* This convenient result occurs because the system equations of motion implicitly relate to system level conservation laws so, taken in reverse, we see global conservation laws constraining local dynamics and local force descriptions such that forces between objects strictly lie along the line joining the objects. This is developed more extensively in the context of Noether's Theorem [26] in a later section. For now, let's consider the center-of-mass system in detail, starting with a description of the center-of-mass coordinate that has equation of motion:

$$\vec{R} = \frac{\sum m_i \vec{r}_i}{\sum m_i}; \quad M = \sum m_i; \quad M\frac{d^2\vec{R}}{dt^2} = \sum_i \vec{F}_i = \vec{F}^{(ext)},$$

where this relates to the equations of motion for the individual objects upon elimination of center-of-mass coordinate:

$$\sum m_i \frac{d^2\vec{r}_i}{dt^2} = \sum_i \vec{F}_i.$$

A direct comparison with the individual equation of motion above, when it's summed over objects, shows that we must have:

$$\sum_{i,j} \vec{F}_{ji} = 0 \rightarrow \vec{F}_{12} = -\vec{F}_{21} \text{ in fundamental case of two objects,}$$

(Eqn. 2-3)

thus, we obtain the weak law of action and reaction (so far). Now let's turn our attention to the system description of angular motion (about the center), which relates to conservation of angular momentum. Starting with the system angular momentum and the change in angular momentum with external torque:

$$L = \sum_i \vec{r}_i \times \vec{p}_i; \quad \frac{dL}{dt} = \sum_i \vec{r}_i \times \vec{F}_i,$$

we first take the time derivative directly:

$$\frac{dL}{dt} = \sum_i \dot{\vec{r}}_i \times \vec{p}_i + \vec{r}_i \times \dot{\vec{p}}_i = \sum_i \vec{r}_i \times \dot{\vec{p}}_i$$

A direct comparison of the time derivatives of the angular momentum then indicates we must have:

$$\sum_{i,j} \vec{r}_i \times \vec{F}_{ji} = 0.$$

(Eqn. 2-4)

Again, let's focus on two objects interacting (labeled 1 and 2): $\vec{r}_1 \times \vec{F}_{21} + \vec{r}_2 \times \vec{F}_{12} = 0$, and since $\vec{F}_{ji} = -\vec{F}_{ij}$ already, we must have: $(\vec{r}_1 -$

13

$\vec{r}_2) \times \vec{F}_{12} = 0$, completing the strong law of action-reaction proof -- the forces strictly lie along the line joining the objects (allowing a potential function description in later analysis).

2.2 D'Alembert's Principle of Virtual Work

This section summarizes D'Alembert's argument in modern notation according to [25,37]. Suppose the system is in equilibrium, i.e., $\vec{F}_i = 0$, then clearly $\vec{F}_i \cdot \delta\vec{r}_i = 0$. So, $\Sigma \vec{F}_i \cdot \delta\vec{r}_i = 0$, which we now decompose as:

$$\vec{F}_i = \vec{F}_i^{(a)} + f_i,$$

(Eqn. 2-5)

where $\vec{F}_i^{(a)}$ is the applied force and f_i is the force of constraint. Thus,

$$\Sigma_i \, \vec{F}_i^{(a)} \cdot \delta\vec{r}_i + \Sigma_i \, \vec{f}_i \cdot \delta\vec{r}_i = 0,$$

where the $\delta\vec{r}_i$ can be arbitrary displacements. We now restrict to the situation where the net virtual work due to the forces of constraint is zero, $\Sigma_i \, \vec{f}_i \cdot \delta\vec{r}_i = 0$, to then get:

$$\Sigma_i \, \vec{F}_i^{(a)} \cdot \delta\vec{r}_i = 0.$$

Suppose the system is now in a general setting, $\vec{F}_i = \dot{\vec{p}}_i$, if we split off the force of constraint as before:

$$\Sigma_i \, \left(\vec{F}_i^{(a)} - \dot{\vec{p}}_i\right) \cdot \delta\vec{r}_i + \Sigma \, \vec{f}_i \cdot \delta\vec{r}_i = 0$$

and, with the same assumption of zero net virtual work due to constraints, we get:

$$\Sigma_i \, \left(\vec{F}_i^{(a)} - \dot{\vec{p}}_i\right) \cdot \delta\vec{r}_i = 0, \qquad D'Alembert's \; principle$$

(Eqn. 2-6)

From the above form we must transform to generalized coordinates that are independent of each other, such that the coefficients of the displacements can be set to zero separately:

$$\vec{r}_i = \vec{r}_i(q_1, q_2, \dots q_n, t) \rightarrow \delta\vec{r}_i = \Sigma_j \frac{d\vec{r}_i}{\partial q_j}\delta q_j \, .$$

First consider the transformation of the $\vec{F}_i^{(a)} \cdot \delta\vec{r}_i$ part (dropping the 'applied' superscript):

$$\Sigma_i \, \vec{F}_i \cdot \delta\vec{r}_i = \Sigma_{i,j}\vec{F}_i \cdot \frac{\partial \vec{r}_i}{\partial q_j}\delta q_j = \Sigma_j \, Q_j \delta q_j$$

$$\rightarrow Q_j = \Sigma_i \, \vec{F}_i \cdot \frac{\partial \vec{r}_i}{\partial q_j}$$

(Eqn. 2-7)

14

where the dimension of Q need not be the dimension of the force, nor the generalized coordinates the dimensions of length, but their product must still be the dimension of work. Now let's consider the transformation of the $\Sigma_i \; \dot{p}_\iota \cdot \delta \vec{r}_i$ term:

$$\Sigma_i \; \dot{p}_\iota \cdot \delta \vec{r}_i = \Sigma_i \; m_i \ddot{\vec{r}} \cdot \delta \vec{r}_i = \Sigma_{i,j} m_i \ddot{\vec{r}} \cdot \frac{\partial \vec{r}_i}{\partial q_j} \delta q_j$$

$$= \Sigma_{i,j} \left\{ \frac{d}{dt} \left(m_i \ddot{\vec{r}} \cdot \frac{\partial \vec{r}_i}{\partial q_j} \right) - m_i \ddot{\vec{r}} \frac{d}{dt} \left(\frac{\partial \vec{r}_i}{\partial q_j} \right) \right\} \delta q_j$$

now,

$$\frac{d}{dt} \left(\frac{\partial \vec{r}_i}{\partial q_j} \right) = \Sigma_k \frac{\partial^2 \vec{r}_i}{\partial q_j \partial q_k} \dot{q}_k + \frac{\partial^2 \vec{r}_i}{\partial q_j \partial t} = \frac{\partial}{\partial q_j} \frac{d\vec{r}_i}{dt} = \frac{\partial \vec{r}_i}{\partial q_j}.$$

Furthermore, switching to $\dot{\vec{r}}_i = \vec{v}_j$:

$$\frac{\partial \vec{v}_\iota}{\partial \dot{q}_j} = \frac{\partial}{\partial \dot{q}_j} \left\{ \Sigma_k \frac{\partial r_i}{\partial q_k} \dot{q}_k + \frac{\partial r_i}{\partial t} \right\} = \frac{\partial r_i}{\partial q_j}$$

We can now write

$$\Sigma_i \; \dot{p}_\iota \cdot \delta \vec{r}_i = \Sigma_i \; \left\{ \frac{d}{dt} \left(m_i \vec{v}_i \cdot \frac{\partial \vec{v}_j}{\partial \dot{q}_j} \right) - m_i \vec{v}_i \cdot \frac{\partial \vec{v}_j}{\partial q_j} \right\}$$

$$= \Sigma_i \; \left\{ \frac{d}{dt} \frac{\partial}{\partial \dot{q}_j} \left(\Sigma_i \; \frac{1}{2} m_i \vec{v}_i^{\;2} \right) - \frac{\partial}{\partial q_j} \left(\Sigma_i \; \frac{1}{2} m_i \vec{v}_i^{\;2} \right) \right\}$$

and writing the kinetic energy term $\Sigma_i \; \frac{1}{2} m_i \vec{v}_i^{\;2} = T$, we get D'Alembert's Principle in the form:

$$\Sigma_j \; \left[\left\{ \frac{d}{dt} \left(\frac{\partial T}{\partial \dot{q}_j} \right) - \frac{\partial T}{\partial q_j} \right\} - Q_j \right] \partial q_j = 0.$$

(Eqn. 2-8)

Using Force written in terms of a potential function, $\vec{F}_i = -\nabla_i V$ (where equipotential surfaces are well-defined in relation to 'field lines'), we have:

$$Q_j = \Sigma_i \; \vec{F}_i \cdot \frac{\partial \vec{r}_i}{\partial q_j} = -\Sigma \nabla_i V \cdot \frac{\partial \vec{r}_i}{\partial q_j} = -\frac{\partial V}{\partial q_j}$$

(Eqn. 2-9)

If we now introduce the standard Lagrangian $L = T - V$, we find that D'Alembert's principle gives rise to the equations of motion expressed in terms of the Lagrangian:

$$\frac{d}{dt} \left(\frac{\partial L}{\partial \dot{q}_j} \right) - \frac{\partial L}{\partial \dot{q}_j} = 0,$$

(Eqn. 2-10)

where the latter succinct form of the equations of motion are known as the Euler-Lagrange (E-L) equations. This completes the derivation of the E-L equations by way of D'Alembert's principle; we will perform a different derivation of the E-L equation in the context of Hamilton's Principle of Least Action in the next chapter.

Let's now consider some of the simplest force fields or phenomenology. Suppose the force acts in a single direction (uniformly) and is constant, such would be an example of the Force due to gravity at the Earth's surface, where $F = -mg$. When taken with the simple pendulum we have a complete description since all other 'system' parameters involve the pendulum (arm length, which is massless, and pendulum bob mass):

Example 2.1. The simple pendulum

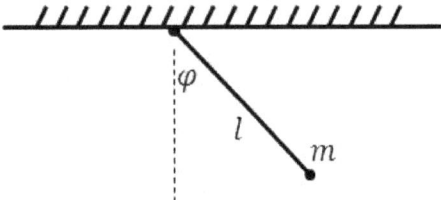

Fig. 2.1 Simple Pendulum.

The Lagrangian is given by $L=KE-PE$ where:

$$KE = \frac{1}{2}m(l\dot{\varphi})^2 \quad and \quad PE = -lgm\cos\varphi, \quad thus \; L$$
$$= \frac{1}{2}m(l\dot{\varphi})^2 + lgm\cos\varphi$$

Exercise 2.1. What are the equations of motion for the simple pendulum?

Example 2.2. The simple spring
Let's now consider where force isn't a constant, but linear in some displacement, such would be the case for a simple spring where $F = -kx$. Here k enters as a phenomenological parameter, not a simple dimensional parameter, and is material dependent. The equations of motion are thus:

16

$$m\ddot{x} = -kx \rightarrow x = \cos(\omega t) + B\sin(\omega t), \qquad where\ \omega = \sqrt{\frac{k}{m}}.$$

Exercise 2.2. What is the Lagrangian?

Example 2.3. The table-spring problem.
Consider a spring with one end attached on a table surface, the other end attached to a mass m. For planar motion in polar coordinates we have for kinetic energy: $T = \left(\frac{1}{2}\right)m(\dot{r}^2 + r^2\dot{\theta}^2)$. For potential energy, from Hooke's Law: $\delta W = -kr\delta r$. The equations of motion then give: $m\ddot{r} - mr\dot{\theta}^2 = -kr$ and $\frac{d}{dt}(mr^2\dot{\theta}) = 0$.

Exercise 2.3. Redo in rectilinear coordinates.

The last example shows how familiarity manipulating differential equations will be helpful in what follows. For this reason a review of Ordinary Differential Equations is given in the appendix (App. A), with a brief overview in what immediately follows for convenience. Then, several more EOM and Lagrangian examples will be given in Sec. 3.3.2, once we've learned how to deal with constraints.

2.3 Overview of simple trajectory-based Ordinary Differential Equations
Some brief comments on the role ordinary differential equations (ODE's) at this early juncture are now given, with more background and numerous examples given in Appendix A. For what follows we are interested in forces that are polynomial in displacement, and at low order, thus ma=F becomes: ma=0; ma=constant; or ma=-kx; as already mentioned. Since $a = \ddot{x}$, we see that we are describing the family of ODE's involving second-order derivatives. Missing from a more general form of such an ODE would be first-order derivative terms, and in adding such we've now included standard frictional forces (if linear in first derivative and negative). Thus we find, almost effortlessly, how added terms in the ODE relate to physics kinematics and phenomenology, and can even be used by such (in reverse) to identify new physical effects, as done by Landau and Lifshits in the discovery of the LL equation [49], and in categorizing various coupling phenomenon [50]. Further analysis of the interplay of ODE's and phenomenology, together with dimensional analysis, is given in Ch. 9.

Chapter 3. Hamilton's Principle of Least Action

We now obtain the Euler-Lagrange equations a different way, as the result of a variational minimum given by Hamilton's Principle of Least Action [10-13]. This approach is more than a Newtonian reformulation as it is the root formulation for the complete quantum theory to be described in [42], and briefly discussed in Sec. 3.2. Thus, this section is of special note in its part of the conceptual foundation for the fully generalized quantum (propagator) theory ([42-44]) and emanator theory ([47]).

3.1 Lagrangian for point-particle

Consider a point-like object and let's define its position by the generalized coordinates $\{q_k\}$, where for K dimensions we have coordinates: q_1 ... q_k ... q_K. Let's now introduce a time parameterization (coordinate) t and define the associated generalized coordinate (position) changes with time, e.g. the velocities. Thus, for coordinates $\{q_k\}$ and velocities $\{v_k\}$ we have:

$$v_k = \frac{dq_k}{dt} = \dot{q}_k,$$

(Eqn. 3-1)

for time t. In early physics it was argued [2-13] that variational constructs that are minimized (such as paths) or maximized (such as entropy), should determine how systems evolve, propagate, or equilibrate. In those discussions we see how the early dynamical description of Newton, $F = ma$, is a second derivative formulation.

The name of the variational function of coordinates and velocities, as before, is the "Lagrangian", and denoted by L:

$$L = L(\{q_k\}, \{\dot{q}_k\}) = L(\{q_k\}, \{v_k\}),$$

where $L = L(\{q_k\}, \{\dot{q}_k\})$ is the form of a preamble that will be often used to indicated the independent variables (variationally relevant) in the function definition, here the coordinates and their velocities. Consider Newton's 2nd Law with no force present, the Lagrangian for this is:

$$L = L(\{q_k\}, \{v_k\}) = \sum_k \frac{1}{2} m(v_k)^2,$$

or, for 1 dimension, have $L=(1/2)mv^2$, the classic expression for kinetic energy. To recover Newton's 2nd Law, we then set the time derivative of each of the Lagrangian velocity derivatives to zero (*not the time derivative of the Lagrangian function itself*):

$$\frac{d}{dt}\frac{dL}{dv} = \frac{d}{dt}\frac{d}{dv}\left(\frac{1}{2}mv^2\right) = m\frac{dv}{dt} = ma = 0,$$

thereby recovering the equation of motion when no Force is present (ma=F=0). Thus, a direct expression of a variation of a function, such that setting that variation to zero yields the equations of motion, is what is obtained in the "action formulation" (first expressed by Hamilton in 1834 with the principle of least action [10-13]). The action S is introduced as a function of a function (a functional) defined by the following integral relation along paths parameterized by time parameter t (see Fig. 2.1):

$$S = \int_{t_1}^{t_2} L(q, \dot{q}, t)dt$$

(Eqn. 3-2)

where the component subscripts are dropped (or one-dimensional case). We will assume this is a valid starting point for deriving equations of motion and we prove this to be the case later in the analysis (where this notion of action is re-derived in the Hamilton-Jacobi formulation in Ch. 8).

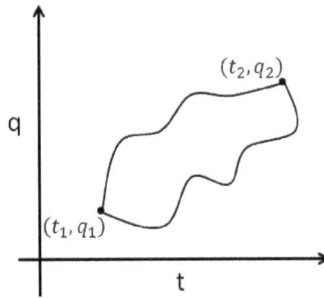

Fig. 3.1. The Action consists of the integration of the Lagrangian along a specified path. Stationarity in the variation of the action, with fixed endpoints, gives rise to the usual Euler Lagrange equations. Two paths of integration for the Lagrangian are shown in the figure, with endpoints shared (fixed) such that $q_1 = q(t_1)$ and $q_2 = q(t_2)$.

In the Hamilton formulation, the motion is given by the time-parameterized path $q(t)$ that gives a stationary value for the action (functional variation is zero), and where the typical boundary conditions is that the endpoints on paths of motion are fixed at beginning t_1 and end t_2, i.e $\delta q(t_1) = \delta q(t_1) = 0$. Assuming that there is not direct time dependence in the Lagrangian, we then have for the functional derivative:

$$0 = \delta S = \delta \int_{t_1}^{t_2} L(q, \dot{q}) dt$$

$$= \int_{t_1}^{t_2} \delta L(q, \dot{q}) dt = \int_{t_1}^{t_2} \left[\left(\frac{\partial L}{\partial q} \right) \delta q + \left(\frac{\partial L}{\partial \dot{q}} \right) \delta \dot{q} \right] dt$$

$$\delta S = \int_{t_1}^{t_2} \left[\left(\frac{\partial L}{\partial q} \right) \delta q + \left(\frac{\partial L}{\partial \dot{q}} \right) \frac{d\delta q}{dt} \right] dt$$

$$= \int_{t_1}^{t_2} \left[\left(\frac{\partial L}{\partial q} \right) \delta q - \frac{d}{dt} \left(\frac{\partial L}{\partial \dot{q}} \right) \delta q + \frac{d}{dt} \left(\frac{\partial L}{\partial \dot{q}} \delta q \right) \right] dt$$

$$\delta S = \left[\frac{\partial L}{\partial \dot{q}} \delta q \right]_{t_1}^{t_2} + \int_{t_1}^{t_2} \left[\left(\frac{\partial L}{\partial q} \right) - \frac{d}{dt} \left(\frac{\partial L}{\partial \dot{q}} \right) \right] \delta q dt$$

The boundary term from the integration by parts is zero since the boundaries are fixed for the variations being considered. This is the standard case for most of the variational problems that will be described. There are alternate, more complex, formulations with non-fixed ends that will be discussed as needed. Thus, we now have that Hamilton's principle of Least Action (standard form) recovers the Euler-Lagrange Equations [8], mentioned previously:

$$\delta S = 0 \Rightarrow \left(\frac{\partial L}{\partial q} \right) - \frac{d}{dt} \left(\frac{\partial L}{\partial \dot{q}} \right) = 0.$$

(Eqn. 3-3)

The Euler-Lagrange equations will be used in the sections that follow to obtain the equations of motion in a huge variety of applications. Before moving to these examples, however, there is more that can be gleaned from the action formulation than a mere recovery of the equations of motion, a variety of properties of motion and conservation laws can now be extracted.

3.1.1 Mechanical properties indicated by action formulation

Prior sections made reference to Goldstein's textbook [25] numerous times, and some of the development (strong law of action-reaction) was from solving problems from there. Going forward we solve in detail many of the problems presented in Landau and Lifshitz's textbook on Mechanics [27], and follow their mathematical development in part as it is an exposition of the possible second–order differential equations that can occur. The ODE-centric approach is also done in Percival's text [28], so this is a popular approach. The role of ordinary differential equations (ODE's) in the development of mechanics is made even more explicit in the effort presented here, however, with a large appendix on ODE's and problems/solutions for such (drawn from notes taken while at Caltech in AMa101, a graduate-level mathematics course on ODE's). Part of the development presented here pairs classes of ODE's with classes of motion, and from there shows how to arrive at general systems, including those with chaos. The chaos part of the discussion is mainly done in the Hamiltonian formulation similar to the textbook by Percival [28]. Advanced dynamics sections draw from problem solutions for problems given in the textbooks by Goldstein [25], Landau and Lifshitz [27],and Fetter & Walecka [29]; and from notes from Dynamics (Ph 106) and Advanced Dynamics (Aph107) courses taken at Caltech (ca. 1986).

Following the description given by Landau and Lifschitz, in Mechanics [27], let's first consider a system comprised of two parts with negligible interaction. We write the total system Lagrangian as the simple addition of their two parts:

$$L = L_1 + L_2.$$

The additive property implies a decoupling of non-interacting systems but with common shared constant (e.g., choice of units). To show this, consider multiplying the Lagrangian by a constant, the resulting equations of motion are unaltered, and the separate terms all share that same multiplier. Continuing along this vein, consider adding a total time derivative of a function (dependent on coordinates and time) to the given definition of a Lagrangian:

$$\tilde{L} = L + \frac{d}{dt}f(q,t)$$

The new action functional obtained is:

$$\tilde{S} = S + f(q(t_2),t_2) - f(q(t_1),t_1)$$

for which the variation is the same when the endpoints are fixed:

$$\delta\tilde{S} = \delta S.$$

Thus, a Lagrangian defines the same equation of motion for any variation if differing by a total time derivative. (If there are non-fixed or non-trivial boundary conditions then there is no longer invariance upon addition of a total time derivative.)

If the Lagrangian is not dependent on spatial coordinate, we say there is homogeneity in space, same for time. If the Lagrangian is not dependent on direction in space, we say there is spatial isotropy, while for time, a 1-dimensional parameter, this is equivalent to saying time reversal invariance. So, if we say there is nothing special about the position or time in describing the free motion of a particle, then we are saying that the Lagrangian for its motion should have no $\{q, t\}$ dependence. Furthermore, the velocity dependence must only depend on the magnitude (for isotropy) which can be conveniently written as a dependence on the magnitude of the velocity squared:

$$L = L(v^2).$$

If this is a valid functional form for the Lagrangian then we expect no change under velocity shift (true for non-relativistic, i.e. Galilean, absolute time reference). Let's try $\vec{v}' = \vec{v} + \vec{\varepsilon}$:

$$L' = L(v'^2) = L(v^2 + 2\vec{v}\cdot\vec{\varepsilon} + \varepsilon^2) = L(v^2) + \frac{\partial L}{\partial v^2}2\vec{v}\cdot\vec{\varepsilon} + O(\varepsilon^2),$$

where the derivation to first order in $\vec{\varepsilon}$ is explicitly shown. For this to remain unaltered at first order, then the first order term must be a total time derivative. Since it already has a time derivative in the velocity, this is only possible if $\frac{\partial L}{\partial v^2}$ is independent of velocity (but nonzero), thus have $L \propto v^2$, and by convention with Newton's specification of mass and inertia we have:

$$L = \frac{1}{2}mv^2,$$

(Eqn. 3-4)

for the free particle, from which application of the Euler-Lagrange equation yields for equation of motion $v = $ constant, recovering the Law of Inertia. Note also that $v^2 = \left(\frac{dl}{dt}\right)^2 = \frac{(dl)^2}{(dt)^2}$, where expressions for the metric, $(dl)^2$, in various coordinate systems are:

Cartesian: $(dl)^2 = (dx)^2 + (dy)^2 + (dz)^2$ $\qquad \Rightarrow L = \frac{1}{2}m(\dot{x}^2 + \dot{y}^2 + \dot{z}^2)$

Cylindrical: $(dl)^2 = (dr)^2 + (r\, d\varphi)^2 + (dz)^2$ $\qquad \Rightarrow L = \frac{1}{2}m(\dot{r}^2 + r^2\dot{\varphi}^2 + \dot{z}^2)$

23

Spherical: $(dl)^2 = (dr)^2 + (r\,d\theta)^2 + (r\,\sin\theta\,d\varphi)^2 \implies L = \frac{1}{2}m(\dot{r}^2 + r^2\dot{\theta}^2 + r^2\sin^2\theta\,\dot{\varphi}^2)$

(Eqn. 3-5abc)

3.1.2 The Action for free motion
Example 3.1. The action for free motion – minimal practical use, maximal theoretical implication

For a free particle with one-dimensional motion we have $L = T = \frac{1}{2}\dot{x}^2$, for which the action is:

$$S = \int_{t_A}^{t_B} L\,dt = \int_{t_A}^{t_B} \frac{1}{2}v^2\,dt,$$

where $v = \frac{x_B - x_A}{t_B - t_A}$ from E-L equation. Thus,

$$S = \frac{1}{2}\frac{(x_B - x_A)^2}{(t_B - t_A)} \quad \rightarrow \quad S = \frac{1}{2}\frac{(\Delta x)^2}{(\Delta t)} \quad \rightarrow \quad (\Delta x)^2 \cong (\Delta t) \; if \; S$$
$$= constant.$$

If $\Delta t = N$ time-steps, then $|\Delta x| \approx \sqrt{\Delta t}$, as with a random walk (further details in [45]).

Exercize 3.1. Repeat with $L = \cosh v$.

Note that the action for free motion is like the solution to the diffusion equation (solution to 1D heat equation), which is our first hint of the possibility of the Schrodinger equation, and the first hint of Ito Integral (Weiner Integral) formulations, seen again later with the Euclideanized quantum form by way of analytic time (via Wick rotation, see [43,44]). The relation to the diffusion relation in one-dimension is also an early hint of the deep connections between dynamics and thermodynamics overall -- via (quantum) mechanics with complex time or analyticity (to be discussed in [43,44]). The reification of analytic trigintaduonion emanation associations or projections, with the emergence of thermality (martingale thermodynamics), geometry (standard cosmology), and gauge geometry (the standard model), is discussed further in [45].

Example 3.2. Lagrangian with higher order time derivatives
Consider a system with the following Lagrangian:

$$L = A\ddot{x}^2 + \frac{1}{2}m\dot{x}^2.$$

24

The equation of motion for such a system can be obtained, uniquely, if we require that the action be an extremum for all paths with the same values of x, and all of its time derivatives, at the endpoints of the paths:

$$S = \int_{t_1}^{t_2} \left(A\ddot{x}^2 + \frac{1}{2}m\dot{x}^2 \right) dt = \int_{t_1}^{t_2} L(\dot{x}, \ddot{x}) dt$$

$$0 = \delta S = \int_{t_1}^{t_2} \left(\frac{\partial L}{\partial \dot{x}} \delta\dot{x} + \frac{\partial L}{\partial \ddot{x}} \delta\ddot{x} \right) dt$$

$$= \int_{t_1}^{t_2} \left(-\frac{d}{dt}\left(\frac{\partial L}{\partial \dot{x}}\right) \delta x - \frac{d}{dt}\left(\frac{\partial L}{\partial \ddot{x}}\right) \delta\dot{x} \right) dt$$

and another integration by parts (with boundary terms dropped, thus total derivatives dropped):

$$\delta S = \int_{t_1}^{t_2} \left(-\frac{d}{dt}\left(\frac{\partial L}{\partial \dot{x}}\right) + \frac{d^2}{dt^2}\left(\frac{\partial L}{\partial \ddot{x}}\right) \right) \delta x dt = 0 \rightarrow \frac{d^2}{dt^2}\left(\frac{\partial L}{\partial \ddot{x}}\right) - \frac{d}{dt}\left(\frac{\partial L}{\partial \dot{x}}\right)$$
$$= 0$$

The equation of motion is thus:
$$2Ax^{(4)} - m\ddot{x} = 0,$$
where the (4) denotes a fourth-order time derivative.

Exercize 3.2. Repeat with $L = A\ddot{x}^3 + \frac{1}{2}m\dot{x}^2 + B\ddot{x}$

3.2 Least Action from highly oscillatory integrals and stationary phase

The variational extremum indicated in Hamilton's principle of least action can also be obtained via an exponentiated large magnitude functional integral [6], where the Action is evaluated along every path each contributing an exponentiated term with a large constant factor (such that a variational minimum dominates, according to negative sign convention below). This is also used in the quantum path integral formulation [48] (and [42]) where there is still a large constant (the inverse of Planck's constant) but the exponentiated term is made imaginary, i.e., each path now contributes its action as a phase term, where stationary phase then selects for the variational extremum. Thus, the classical integral form can be analytically continued into a quantum integral form that is directly relevant:

$$\int e^{-Mf(x)} dx \quad \rightarrow \quad \int e^{iMf(x)} dx, \quad where \; M \gg 1.$$

(Eqn. 3-6)

Note that the classical integral form was an odd representation, not much used since it reduced back to Hamilton's least action anyway. In its

25

complex form, however, when reduced to differential form consistent with least action we get Schrodinger's equation, and recover the classical theory at lowest order, with quantum corrections at higher order (see [42] for details).

The notion of multiple paths, from which the path imparting stationarity is selected, is foundational with the quantum PI approach to quantum mechanics. PI quantization is equivalent in various domains to the operator/wavefunction (Schrodinger) or self-adjoint operator/Hilbert Space (Heisenberg) formulations, as will be shown in [42], where the choice of formulation to solve a problem can be critical to its solution. The variationally-defined classical constructs, especially those outlined in Ch. 8, will eventually generalize to the full quantum mechanical formulation (in terms of multiple propagation paths and a stationary action functional over those paths). In practice, the full quantum theory, especially for bound systems, is much easier to analyze if we shift from the path integral representation to one of the equivalent formulations by Heisenberg [16], Schrodinger [17], or Dirac [18], as will be shown in [42]. The Heisenberg operator-calculus formulation is based on an operator reformulation of the classical Hamiltonian (Ch. 6); Schrodinger's equation is based on an operator-wavefunctional reformulation of the Hamilton-Jacobi equations (Ch. 8); and Dirac's axiomatic reformulation [42] shifts to general systems without necessarily having a classical analogue (and also bridges to the relativistic wave-equation for spin ½ fermions in further developments [18]).

Notice that the classical integral representation involved a simple sum on paths (no weighting) and later, with analytic continuation to a quantum formulation, we still had a sum on paths that was unweighted. This characteristic is carried over to statistical mechanics to become the equipartition theorem, and can be found via analytic continuation (Wick rotation) from the quantum propagator to the statistical mechanical partition function (described in Books 7 and 8 of the Series). Thus, there is a growing body of evidence that the underlying theories, or theoretical representations, are analytic, and possibly in multiple ways, indicating possibly being fundamentally hypercomplex (discussed further in Book 9).

3.3 Lagrangian for system of particles

Now consider a bunch of freely moving particles, the Lagrangian consists of kinetic energy terms:

$$L = T = \sum_a \frac{1}{2} m_a \, v_a{}^2,$$

where the index 'a' ranges over the different particles, with the Lagrangian for one-dimensional motion explicit. Multi-dimensional motion (typically three dimensional) is implicit where component indices on vector quantities are suppressed. Let's now consider the particles to be interacting and express this as a "potential energy" term as indicated from the previous D'Alembert/Newtonian formulation:

$$L = \sum_{a=1} \frac{1}{2} m_a \, v_a{}^2 - U(\vec{r}_1, \vec{r}_2, \ldots) = T - U,$$

(Eqn. 3-8)

where the standard notation "T" for kinetic energy and "U" for potential energy has been introduced. The Euler-Lagrange equations, using standard vector notation explicitly on velocities, then yield:

$$m_a \frac{d\vec{v}_a}{dt} = -\frac{\partial U}{\partial \vec{r}_a} = \vec{F},$$

(Eqn. 3-9)

where F is the familiar Newtonian force. Notice to arrive at this from the Lagrangian we once again see introduction of a potential function with no reference to time or to information transmission, e.g., it references an implicit Galilean absolute time, with instantaneous propagation of interactions. Obviously this will begin to err significantly when velocities become relativistic, but at this stage, where we examine classical mechanical properties in classical settings (such as pendulum motion), this is a negligible error. Recall that the Lagrangian is unchanged to within an additive constant or a total time-derivative. So far we aren't considering potentials with time dependence so focusing on the "unchanged to within an additive constant" means we are free to shift our Lagrangian formulation as convenient to have the potential fall to zero as the distance between particles grows large

Let's now consider a system of two particles as seen from the viewpoint of a system defined in terms of the first particle (now seen as an open system). First, the Lagrangian for just two particles is:
$$L = T_1(q_1, \dot{q}_1) + T_2(q_2, \dot{q}_2) - U(q_1, q_2).$$

Suppose we have a solution for the second particle as a function of time: $q_2 = q_2(t)$, and that we substitute this solution back into our Lagrangian. What results is a kinetic term where the only independent variable is now time, thus can be viewed as a total time derivative, and thus dropped from the Lagrangian without altering its equations of motion. The equivalent Lagrangian, where now the first particle is described in an "open" system is thus:

$$L = T_1(q_1, \dot{q}_1) - U(q_1, q_2(t)).$$

The Lagrangian has now arrived at its main form $L = T - U$, kinetic energy minus potential energy. It might seem odd at this point to have a fundamental entity $T - U$ in the variational formalism, when conservation of overall energy would govern $T + U$. (It turns out the latter works as a basis for a variational, Hamiltonian, formalism as well, that we will get to in later chapters.) For now, we stay with the Lagrangian formulation and move to the type of "potential" implicit in a system by way of constraints.

3.3.1 Constraints

Mechanical systems often deal with motion under constraint by means of rods, strings, hinges. Two new issues then arise: (1) determining the effect of constraint on the degrees of freedom (N particles in 3D have 3N degrees of freedom while unconstrained, if forced onto a surface, for example, then reduced to 2N degrees of freedom, etc.); and (2) friction. In the following example problems we assume friction is negligible, but return to a discussion of friction and other phenomenological forces in Ch. 9.

If a constraint is nonholonomic the equations expressing the constraint cannot be used to eliminate the dependent coordinates. Consider general linear differential equations of constraint of the form:

$$\sum_{i=1}^{n} g_i(x_1, \dots, x_n) dx_i = 0.$$

Constraints can often be put in this form but it is integrable (and holonomic) only if there exists an integrating function $f(x_1, \dots, x_n)$:

$$\frac{\partial(f g_i)}{\partial x_j} = \frac{\partial(f g_j)}{\partial x_i}.$$

Thus, the second-order mixed derivatives of an integrable function should not depend on order of differentiation. As an example of this, consider a disk rolling in a plane, with constraint governed by a pair of differential equations (with explicit zero factors shown):

$$0 d\theta + dx - a \sin\theta \, d\varphi = 0 \quad \text{and} \quad 0 d\theta + dy + a \cos\theta \, d\varphi = 0.$$

For this we have:
$$\frac{\partial(f(1))}{\partial\theta} = \frac{\partial(f(0))}{\partial x} = 0 \quad\rightarrow\quad \frac{\partial f}{\partial\theta} = 0$$
$$\rightarrow \quad f \text{ has no } \theta \text{ dependance.}$$

But this is inconsistent with:
$$\frac{\partial(f(1))}{\partial\varphi} = \frac{\partial(f(-a\sin\theta))}{\partial x} \quad\rightarrow\quad f \text{ has } \theta \text{ dependance.}$$

Thus, rolling objects are a familiar example of a system with constraints that are nonholonomic.

3.3.2 Lagrangians for simple systems

If there are simple constraints or couplings, direct evaluation of kinetic terms is possible. Consider the simplest double pendulum for example (shown in Fig. 3.2, made from massless rods joining point-masses). Note that general multi-element systems will almost entirely be covered in [44] on Statistical Mechanics.

Example 3.3 The double pendulum

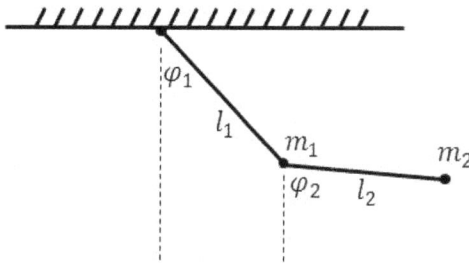

Fig. 3.2. The double pendulum.

Let's describe the coordinates of the m_2 mass by (x,y):
$$x = l_1\sin\varphi_1 + l_2\sin\varphi_2 \quad\text{and}\quad y = l_1\cos\varphi_1 + l_2\cos\varphi_2$$
Then, taking the Lagrangian as kinetic energy minus potential energy, $L = K.E.-P.E.$, we first determine K.E.:
$$K.E. = \frac{1}{2}m_1(l_1\dot\varphi_1)^2$$
$$+ \frac{1}{2}m_2[(l_1\cos\varphi_1\dot\varphi_1 + l_2\cos\varphi_2\dot\varphi_2)^2$$
$$+ (-l_1\sin\varphi_1\dot\varphi_1 - l_2\sin\varphi_2\dot\varphi_2)^2]$$
$$= \frac{1}{2}(m_1 + m_2)(l_1\dot\varphi_1)^2 + \frac{1}{2}m_2(l_2\dot\varphi_2)^2$$
$$+ m_2(l_1\dot\varphi_1)(l_2\dot\varphi_2)\cos(\varphi_1 - \varphi_2)$$
$$P.E. = (m_1 + m_2)g(\sin\varphi_1)l_1 + m_2gl_2\sin\varphi_2$$

29

and the Lagrangian is thus:

$$L = \frac{1}{2}(m_1 + m_2)(l_1\dot{\varphi}_1)^2 + \frac{1}{2}m_2(l_1\dot{\varphi}_1)^2 + m_2(l_1\dot{\varphi}_1)(l_2\dot{\varphi}_2)\cos(\varphi_1 - \varphi_2)$$
$$-(m_1 + m_2)gl_1\sin\varphi_1 - m_2gl_2\sin\varphi_2$$

Exercise 3.3. Determine the equations of motion.

Let's now consider the effect on a simple pendulum of modulating the support point in various ways (horizontal in Ex. 3.4; vertical in Ex. 3.5; and circular in Ex. 3.6):

Example 3.4. The single pendulum with horizontally oscillating support
Let's now consider the single pendulum (Fig. 3.3) when the point of support is now at m_1 and it is oscillating horizontally:

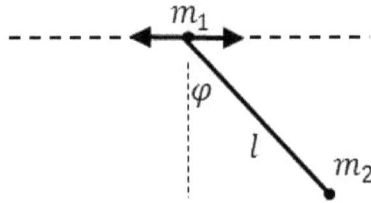

Fig. 3.3. The single pendulum with horizontally oscillating support.

Specifying the second mass carefully in terms of Cartesian coordinates we have:

$$x_2 = x_1 + l\sin\varphi \quad and \quad y_2 = l\cos\varphi.$$

Then, defining the Lagrangian by $L = K.E. - P.E.$ we have:

$$K.E. = \frac{1}{2}m_1\dot{x}_1{}^2 + \frac{1}{2}m_2[(\dot{x}_1 + l\cos\varphi\dot{\varphi})^2 + (-l\sin\varphi\dot{\varphi})^2]$$
$$= \frac{1}{2}m_1\dot{x}_1{}^2 + \frac{1}{2}m_2[\dot{x}_1{}^2 + (l\dot{\varphi})^2 + 2l\cos\varphi\dot{x}_1\dot{\varphi}]$$
$$= \frac{1}{2}(m_1 + m_2)\dot{x}_1{}^2 + \frac{1}{2}m_2(l\dot{\varphi})^2 + m_2l\cos\varphi\dot{x}_1\dot{\varphi}$$

$$P.E. = -lgm_2\cos\varphi$$

$$L = \frac{1}{2}(m_1 + m_2)\dot{x}_1{}^2 + \frac{1}{2}m_2(l\dot{\varphi})^2 + m_2l\cos\varphi(\dot{x}_1\dot{\varphi}$$
$$+ gl)$$

Exercise 3.4. Determine the equations of motion.

Example 3.5. Single pendulum with vertically oscillating support.
Consider Fig. 3.3, but with *vertically* oscillating support. Specifying the
second mass in terms of Cartesian coordinates we have:
$$x_2 = x_1 + l\sin\varphi \quad and \quad y_2 = l\cos\varphi.$$

Then, defining the Lagrangian by $L = K.E. - P.E.$ we have:

$$K.E. = \frac{1}{2}m_1\dot{x}_1{}^2$$
$$+\frac{1}{2}m_2[(\dot{x}_1 + l\cos\varphi\dot{\varphi})^2$$
$$+ (-l\sin\varphi\dot{\varphi})^2]$$
$$= \frac{1}{2}m_1\dot{x}_1{}^2 + \frac{1}{2}m_2[\dot{x}_1{}^2 + (l\dot{\varphi})^2 + 2l\cos\varphi\dot{x}_1\dot{\varphi}]$$
$$= \frac{1}{2}(m_1 + m_2)\dot{x}_1{}^2 + \frac{1}{2}m_2(l\dot{\varphi})^2 + m_2l\cos\varphi\dot{x}_1\dot{\varphi}$$
$$P.E. = -lgm_2\cos\varphi$$
$$L = \frac{1}{2}(m_1 + m_2)\dot{x}_1{}^2 + \frac{1}{2}m_2(l\dot{\varphi})^2 + m_2l\cos\varphi(\dot{x}_1\dot{\varphi}$$
$$+ gl)$$

Exercise 3.5. Determine the equations of motion.
Example 3.6. The single pendulum with rotating disc (oscillating)
support.
Consider Fig. 3.3, but with *rotating disc* oscillating support. Starting with
the coordinates of the pendulum mass:
$$x = l\sin\varphi + a\sin\gamma t \quad and \quad y = l\cos\varphi + a\cos\gamma t.$$
The kinetic energy is then:
$$K.E. = \frac{1}{2}m([l\cos\varphi\dot{\varphi} + a\gamma\cos\gamma t]^2$$
$$+ [-l\sin\varphi\dot{\varphi} + a\gamma\sin\gamma t]^2)$$
$$= \frac{1}{2}m(l\dot{\varphi})^2 + m\gamma al\dot{\varphi}[\cos\varphi\cos\gamma t + \sin\varphi\sin\gamma t]$$
$$= \frac{1}{2}m(l\dot{\varphi})^2 + m\gamma al\dot{\varphi}(\cos(\varphi - \gamma t))$$

and the potential energy is:
$$P.E. = -gml\cos\varphi + gma\cos\gamma t$$
$$L = \frac{1}{2}m(l\dot{\varphi})^2 + m\gamma al\dot{\varphi}(\cos(\varphi - \gamma t)) + gm(l\cos\varphi - a\cos\gamma t)$$
$$= \frac{1}{2}m(l\dot{\varphi})^2 + mla\gamma^2\sin(\varphi - \gamma t) + mgl\cos\varphi$$

Exercise 3.6. Determine the equations of motion.
Let's now consider when the pendulum arm is a spring (see Fig. 3.4).

Example 3.7 The single pendulum with spring for pendulum arm support.

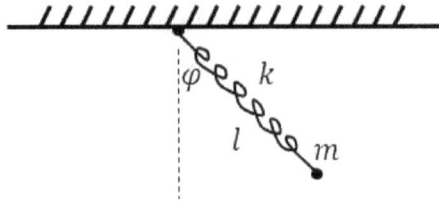

Fig. 3.4. The single pendulum with spring for pendulum arm support.

$$L = \frac{1}{2}m(\dot{r}^2 + r^2\dot{\theta}^2) + mgr\cos\theta - \frac{1}{2}k(r-l)^2$$

$$\frac{d}{dt}\left(\frac{\partial L}{\partial \dot{r}}\right) - \frac{\partial L}{\partial r} = m\ddot{r} - mg\cos\theta + k(r-l)$$
$$+ mr\dot{\theta}^2 = 0$$

$$\frac{d}{dt}\left(\frac{\partial L}{\partial \dot{\theta}}\right) - \frac{\partial L}{\partial \theta} = mr^2\ddot{\theta} + mgr\sin\theta = 0$$

Let's consider small oscillations due to the spring such that the arm length can be written as $r = l + \varepsilon$ with $\varepsilon \ll l$ and taking small angle of oscillation as well, we can write a small oscillation result and identify resonant frequencies (this is an example of a simple small oscillation analysis, with a more extensive description for more complex small oscillation analysis given in Sec. 3.8). To first order we have:
$$m\ddot{\varepsilon} - mg + k\varepsilon = 0 \quad \text{and} \quad ml^2\ddot{\theta} + mgl\theta = 0.$$
Thus, have small oscillation solutions:

$$\varepsilon = A\cos\left(\omega_0^{(1)}t + \alpha\right) + \frac{mg}{k} \quad \rightarrow \quad \omega_0^{(1)} = \sqrt{\frac{k}{m}}$$

and

$$\theta = B\cos\left(\omega_0^{(2)}t + \beta\right) \rightarrow \quad \omega_0^{(2)} = \sqrt{\frac{g}{l}}.$$

Exercise 3.7. What happens if $\omega_0^{(1)} = \omega_0^{(2)}$.

Let's now consider when the pendulum arm can support tension but not compression (e.g., it's a rope).

Example 3.8. The single pendulum with tension-only support for pendulum mass.
Consider Fig. 3.4, but with *tension* support. Again we have the simple pendulum, with mass m held by a string (or wire) of length l, and we now

consider the tension in the wire. We'd like to examine the holonomic regime where the string tension does not go slack. Again we have in polar coordinates, for potential $U = -mgr\cos\theta$:

$$L = \frac{1}{2}m(\dot{r}^2 + r^2\dot{\theta}^2) + mgl\cos\theta$$

Thus

$$E_T = \frac{1}{2}ml^2\dot{\theta}^2 - mgl\cos\theta$$

where the effective force acting on the wire is radial. Let's use the E-L equation for coordinate r:

$$\frac{d}{dt}\left(\frac{\partial L}{\partial \dot{r}}\right) - \frac{\partial L}{\partial r} = Q_r$$

(Eqn. 3-10)

Since $Q_r = -T_r$, the string tension, we then have:

$$m\ddot{r} - mr\dot{\theta}^2 - mg\cos\theta = -T_r \quad \rightarrow \quad T_r = \frac{2}{l}E_T + 3mg\cos\theta$$

$$0 \leq \frac{2}{l}E_T + 3mg\cos\theta \quad \rightarrow \quad E_T$$

$$\geq -\frac{3}{2}mgl\cos\theta \quad \text{for a taut string or rope.}$$

If there exists a maximum angle, θ_{max}, we have:

$$E_T = -mgl\cos\theta_{max} \quad \text{and} \quad 0 \leq \frac{2}{l}E_T + 3mg\cos\theta_{max}$$

$$0 \leq -2mg\cos\theta_{max} + 3mg\cos\theta_{max} \quad \rightarrow \quad 0 \leq \cos\theta_{max} \quad \rightarrow \quad 0 \leq \theta_{max} \leq 90$$

So, if there is a maximum angle for the motion with wire taut it must lie in $0 \leq \theta_{max} \leq 90$, with system energy:

$$-mgl \leq E_T \leq 0.$$

If there is no maximum angle with tautness, then we are meeting the condition $E_T \geq -\frac{3}{2}mgl\cos\theta$ for any angle, thus have:

$$E_T \geq \frac{3}{2}mgl$$

Let's now shift the potential energy such that the pendulum at rest has $E = 0$, then the range of energy values where string tension is maintained is:

$$0 \leq E_T < mgl \quad \text{and} \quad \frac{5}{2}mgl \leq E_T < \infty.$$

Exercise 3.8. How to go from libration to rotation?

Example 3.9. A pendulum with horizontal support motion with spring restoring force.
Let's consider the problem of a pendulum free to move in the horizontal direction whose support point is also free to move in the horizontal direction with spring constant $k/2$ on both left and right sides (similar to problem 3.7 in [29]). The pendulum bob has mass m connected by massless rod of length l to the support point. The bob motion is constrained to lie in a vertical plane of pendulum motion, where we take the coordinates to be:

$$X = x + l \sin \theta \quad and \quad Y = -l \cos \theta$$

The Lagrangian is then:

$$L = \frac{1}{2} m (\dot{X}^2 + \dot{Y}^2) - U, \quad where \ U = \frac{1}{2} kx^2 - mgl \cos \theta$$

which simplifies to:

$$L(x, \theta) = \frac{1}{2} m \dot{x}^2 + \frac{1}{2} m (l \dot{\theta})^2 + m \dot{x} \dot{\theta} l \cos \theta - U.$$

The E-L equation for x gives:

$$m \ddot{x} + \frac{d}{dt} (m \dot{\theta} l \cos \theta) - kx = 0$$

and the E-L equation for θ gives:

$$ml^2 \ddot{\theta} + \frac{d}{dt} (m \dot{x} l \cos \theta) + m \dot{x} \dot{\theta} l \sin \theta + mgl \sin \theta = 0.$$

In the small oscillation approximation, the equations of motion reduce to:

$$\ddot{x} + l \ddot{\theta} - \frac{k}{m} x = 0 \quad and \quad \ddot{x} + l \ddot{\theta} + g \theta = 0.$$

We can combine to see a relation between (x, θ): $x = \frac{mg}{k} \theta$, which reduces to a single relation:

$$L \ddot{\theta} + g \theta = 0 \quad where \ L = l + \frac{mg}{k}.$$

Thus, for small oscillation, we have a pendulum of effective length $L = l + \frac{mg}{k}$.

Exercise 3.9. Redo with mass M for rod (uniform).

Example 3.10. How high can you swing before support tension goes to zero?
The two dynamical systems considered next have identical Lagrangians aside from a shift in angular coordinate. Both have the same constraint of constant radial distance, where the force of constraint going to zero either

34

marks where a pendulum string tension goes slack, or when a sliding object leaves a hemispherically domed surface. Let's consider the pendulum problem first and address the issue of when the pendulum string tension goes to zero.

The first problem also answers the question of whether can you get on a swing and swing in greater and greater arcs, parametrically driven perhaps, and arrive at sufficient angular speed to start doing complete rotations.... The answer is never, because an angular velocity (at bottom of arc) that has $\omega > \sqrt{(5g/l)}$ would be required, with a 'jump' or impulse required since once the angular velocity grows to $\omega = \sqrt{(2g/l)}$ the support line tension goes to zero, and further (incremental or adiabatic) growth in system energy will not be possible.

The Lagrangian for the pendulum is now written with explicit Lagrange multiplier τ (see note below) for the pendulum radius r constrained to be length l:

$$L = \frac{1}{2}m(\dot{r}^2 + r^2\dot{\theta}^2) + mgr\cos\theta - \tau(r - l)$$

The E-L equations give us the equations of motion:

$$r: \quad m\ddot{r} - mr\dot{\theta}^2 - mg\cos\theta - \tau = 0$$

$$\theta: \quad \frac{d}{dt}(mr^2\dot{\theta}) + mgr\sin\theta = 0$$

$$\tau: \quad r - l = 0$$

Note the introduction of a "Lagrange multiplier" such that when treated as a variational parameter in its own right, with its own E-L equation (shown above), where it recovers the constraint equation. Use of Lagrange multipliers in what follows will, similarly, be very simple, where we obtain, for example, a term $-\tau(contraint\_body)$, whenever the constraint equation is $contraint\_body = 0$ (obviously this only works for equality constraints, but there is a very similar procedure for inequality constraints as well [24]).

From the θ equation we get a constant of the motion (conservation of energy):

$$\frac{d}{dt}\left(\frac{1}{2}\dot{\theta}^2 - \frac{g}{l}\cos\theta\right) = 0$$

If we define $\dot{\theta} = \omega$ at $\theta = 0$:

$$\frac{1}{2}\dot{\theta}^2 - \frac{g}{l}\cos\theta = \frac{1}{2}\omega^2 - \frac{g}{l}$$

35

Solving for the tension τ:
$$\tau = ml\omega^2 - 2mg + 3mg\cos\theta$$

Consider when tension (or the force of constraint) goes to zero:
$$\omega^2 = \frac{g}{l}(2 - 3\cos\theta).$$

We see that zero tension solutions exist when $\frac{g}{l}(2 - 3\cos\theta) \geq 0$. The angle at which zero constraint first occurs is for:
$$\cos\theta = \frac{2}{3} \quad \rightarrow \quad \theta \cong 48°.$$

There are three domains of interest in the energy formula:

Case 1: $l\omega^2 < 2g$: $2mg\cos\theta = ml\dot\theta^2 - ml\omega^2 + 2mg > -2mg + 2mg = 0$. Thus, we have $\cos\theta > 0$, thus $\theta \leq 45°$ and since less than $\theta \cong 48°$, the tension $\tau > 0$.

Case 2: $2g < l\omega^2 < 5g$: $2mg\cos\theta = ml\dot\theta^2 - (x - 2)mg$, where $2 < x < 5$. Thus, can have $\tau = 0$ when $\cos\theta = \frac{2}{3} - \frac{l\omega^2}{3g}$ as already noted.

Case 3: $l\omega^2 > 5g$: $\omega^2 = \frac{g}{l}(2 - 3\cos\theta)$ can never be satisfied, thus tension never goes to zero -- the pendulum rotates (completely), rather than librates.

Exercise 3.10. Describe the motion as you go from $l\omega^2 > 5g$ and decrease ω.

Example 3.11. Motion on the surface of a hemisphere
For the second, related, problem consider the motion of a disc (hockey puck) on the surface of a hemisphere. We'd like to know at what angle the sliding disc departs the hemisphere as it slides, e.g., when is the force of constraint zero. The Lagrangian is
$$L = \frac{1}{2}m(\dot{r}^2 + r^2\dot\theta^2) - mgr\cos\theta - \tau(r - l),$$
and the analysis proceeds as before, with the same result for the angle at which the constraint first reaches zero ($\theta \cong 48°$) as before.

Exercise 3.11. What spring constant k, for spring restoring to the top of the hemisphere, will maintain constraint contact up to $\theta = 50°$

3.4 Conserved quantities in Simple Systems

The Hamiltonian for a simple system of particles is described next (typically one element or small group of elements (two) linked somehow), but only in context of identifying integrals of the motion, such as conservation of energy, momentum, and angular momentum. Further discussion of Hamiltonians is then done in Ch. 6.

Consider a generalized coordinate system q_i, where 'i' is the component in a system with s degrees of freedom (the particles' cumulative dimensions of free motion are all counted towards s). Likewise for the associated velocities: $\dot{q}_i$. There are thus s degrees of freedom for the generalized coordinate and s degrees of freedom for the generalized velocity. This gives rise to 2s initial conditions to specify the motion. In a closed mechanical system this would appear to indicate 2s conditions and associated constants or integrals of motion, but the appearance of time in the velocity as a differential means t and $t + t_0$ have the same equation of motion, so one of these 2s constants is merely t_0, a choice of time origin. Let's consider symmetries of the space of motion and implications given the Lagrangian formulation:

$$\frac{dL(q_i, \dot{q}_i, t)}{dt} = \sum_i \left[\left(\frac{\partial L}{\partial q_i} \right) \dot{q}_i + \left(\frac{\partial L}{\partial \dot{q}_i} \right) \ddot{q}_i \right] + \frac{\partial L}{\partial t}$$

First consider homogeneity in time, which means closed system or open system but with time-independent external field. Either way, have $\frac{\partial L}{\partial t} = 0$, and with re-use of the Euler-Lagrange relations:

$$\frac{dL}{dt} = \sum_i \left[\left(\frac{\partial L}{\partial q_i} \right) \dot{q}_i + \left(\frac{\partial L}{\partial \dot{q}_i} \right) \ddot{q}_i \right] = \sum_i \left[\dot{q}_i \frac{d}{dt} \left(\frac{\partial L}{\partial \dot{q}_i} \right) + \left(\frac{\partial L}{\partial \dot{q}_i} \right) \ddot{q}_i \right]$$

$$= \sum_i \left[\frac{d}{dt} \left(\dot{q}_i \frac{\partial L}{\partial \dot{q}_i} \right) \right]$$

Thus,

$$\frac{d}{dt} \left[\sum_i \left(\dot{q}_i \frac{\partial L}{\partial \dot{q}_i} \right) - L \right] = 0$$

The conserved quantity with time is energy, denoted by E:

$$E = \sum_i \left(\dot{q}_i \frac{\partial L}{\partial \dot{q}_i} \right) - L$$

(Eqn. 3-11)

Note that additivity of Energy on subsystems then follows from additivity for the Lagrangian and the explicit additivity indicated by the sum. If $L = T(q, \dot{q}) - U(q)$ and $T(q, \dot{q}) \propto (\dot{q})^2$, which is typical, then *the standard*

37

energy conservation in the form of kinetic energy plus potential energy results:

$$E = T(q, \dot{q}) + U(q).$$

(Eqn. 3-12)

Next consider homogeneity in space, and start from a variational expression on the Lagrangian assumed not explicitly time-dependent:

$$\delta L(q, \dot{q}) = \sum_i \left[\left(\frac{\partial L}{\partial q_i} \right) \delta q_i + \left(\frac{\partial L}{\partial \dot{q}_i} \right) \delta \dot{q}_i \right]$$

where an infinitesimal displacement should not alter the evaluation of the Lagrangian when $\delta q_i \neq 0$:

$$\delta L(q, \dot{q}) = 0 = \sum_i \left(\frac{\partial L}{\partial q_i} \right) = \sum_i - \left(\frac{\partial U}{\partial q_i} \right) \Rightarrow \sum_i F_i = 0.$$

Net Forces and moments on a closed system sum to zero (specialized use of this will be shown in in Sec. 5.1). If we substitute back the Euler-Lagrange relation to get an explicit total time derivative term:

$$\sum_i \frac{d}{dt} \left(\frac{\partial L}{\partial \dot{q}_i} \right) = \frac{d}{dt} \sum_i \left(\frac{\partial L}{\partial \dot{q}_i} \right) = 0 .$$

From the total time derivative relation we get a constant of the motion corresponding to conservation of momentum:

$$\sum_i \left(\frac{\partial L}{\partial \dot{q}_i} \right) = \vec{P} ,$$

(Eqn. 3-13)

where for systems with $T(q, \dot{q}) \propto (\dot{q})^2$ for each of the particles this simplifies to the standard form:

$$\vec{P} = \sum_i m_i v_i .$$

(Eqn. 3-14)

Note: with two particles we have $\vec{F}_1 + \vec{F}_2 = 0$, which is equivalent to saying action equals reaction (i.e. Newton's 3rd law is a special case of conservation of momentum and Lagrange's equation).

To go with our generalized coordinates and velocities, the generalized momenta and forces are:

$$p_i = \frac{\partial L}{\partial \dot{q}_i} \quad and \quad F_i = \frac{\partial L}{\partial q_i},$$

(Eqn. 3-15)

where Lagrange's equations are simply:

$$\dot{p}_i = F_i.$$

(Eqn. 3-16)

38

Now let's see what happens due to isotropy of space. For this we shift from generalized coordinates to a three-dimensional radial position vector with infinitesimal rotational displacement given by:
$$\delta\vec{r} = \delta\vec{\varphi} \times \vec{r} \ and \ \delta\vec{v} = \delta\vec{\varphi} \times \vec{v}.$$
The variation in the Lagrangian should be zero (now indexing over individual particles):
$$0 = \delta L(\vec{r}_a, \dot{\vec{r}}_a) = \delta L(\vec{r}_a, \vec{v}_a) = \sum_a \left[\left(\frac{\partial L}{\partial \vec{r}_a} \right) \cdot \delta\vec{r}_a + \left(\frac{\partial L}{\partial \vec{v}_a} \right) \cdot \delta\vec{v}_a \right]$$
Substituting the E-L equation and definition of generalized momentum:

$$\sum_a \left[\dot{\vec{p}}_a \cdot \delta\vec{r}_a + \vec{p}_a \cdot \delta\vec{v}_a \right] = 0 \implies \delta\vec{\varphi} \cdot \sum_a \left[\vec{r}_a \times \dot{\vec{p}}_a + \vec{v}_a \times \vec{p}_a \right]$$

Thus, arrive at:
$$\frac{d}{dt} \left[\sum_a \vec{r}_a \times \vec{p}_a \right] = 0 \implies \vec{M} = \sum_a \vec{r}_a \times \vec{p}_a = constant.$$

$$\text{(Eqn. 3-17)}$$

The quantity $\vec{M}$ is the angular momentum, and it is conserved. There are no other additive integrals of the motion (e.g., no other global spatial symmetries than homogeneity and isotropy of space).

Now that we know angular momentum is conserved we can begin to explore the ramifications of this. Angular momentum in 1D is trivially zero, thus we must move to problems with 2D unconstrained motion or 3D motion. Let's start with the *spherical* pendulum.

Example 3.12. The spherical pendulum.
Consider Fig. 3.4, but with *tension* support and with motion of mass allowed in 3-D (e.g., no longer horizontally planar). The Cartesian coordinate of the mass is:
$$x = lsin\varphi cos\theta \quad and \quad y = lsin\varphi sin\theta \quad and \quad z = lcos\varphi$$
Their time derivatives are straightforward:
$$\dot{x} = lcos\varphi\dot{\varphi} \ cos\theta + lsin\varphi(-sin\theta)\dot{\theta}, \ etc.$$
The Lagrangian is thus
$$L = \frac{1}{2}m\{l^2(cos^2\varphi\dot{\varphi}^2) + l^2sin^2\varphi\dot{\varphi}^2 + l^2sin^2\varphi\dot{\theta}\}$$
$$- mglcos\varphi$$
$$= \frac{1}{2}m(l\dot{\varphi})^2 + \frac{1}{2}m\left(lsin\varphi\dot{\theta}\right)^2 - mglcos\varphi$$

39

For the equations of motion we start by elimination of the conserved angular momentum about the z-axis:

$$\frac{d}{dt}\left(\frac{\partial L}{\partial \dot{\theta}}\right) - \frac{\partial L}{\partial \theta} = 0 \rightarrow \frac{d}{dt}\left(ml^2 \sin^2\varphi \dot{\theta}\right) = 0$$

$$ml^2 \sin^2\varphi \dot{\theta} = P_\theta \text{ , a conserved quantity, alternatibvely} \Rightarrow \dot{\theta}$$

$$= \frac{P_\theta}{ml^2 \sin^2\varphi}$$

Eliminating the $\dot{\theta}$ dependence in the Lagrangian by use of its conserved quantity we then get the revised Lagrangian:

$$L = \frac{1}{2}m(l\dot{\varphi})^2 + \frac{P_\theta{}^2}{2ml^2\sin^2\varphi} - mglcos\varphi$$

where now:

$$\frac{d}{dt}\left(\frac{\partial L}{\partial \dot{\varphi}}\right) - \frac{\partial L}{\partial \varphi} = 0 \Rightarrow ml^2\ddot{\varphi} = \frac{-P_\theta{}^2 \sin\varphi\cos\varphi}{ml^2\sin^4\varphi} + mglsin\varphi$$

thus,

$$\ddot{\varphi} + \frac{P_\theta{}^2}{(ml)^2}\frac{\cos\varphi}{\sin^3\varphi} - \frac{g}{l}sin\varphi = 0$$

Exercise 3.12. What is the natural frequency in the small angle approximation?

Example 3.13. Table with a hole, threaded by a line with masses at the ends.
Let's consider another scenario where the angular momentum about a particular axis is conserved. Consider a table with a hole. A tension line threads the hole. The end of the line hanging under the table has mass m_2 attached (the line has negligible mass), while the end resting on the table top has mass m. The initial force balance equations provide:

$$F_2 = m_2 g - T_2, \qquad T_2 = T_1 = F_1 = ma_1, \qquad y_2 = l - r_1,$$
$$\dot{y}_2 = -\dot{r}_1, \qquad \ddot{y}_2 = -\ddot{r}_1$$

While the force, in terms of the potential function, provide:

$$F_i = -\frac{\partial U}{\partial q_i}, \quad F_1 = m_1 a_1 = m_1\left(\ddot{r}_1 + r_1{}^2\ddot{\theta}\right) = m_1\ddot{r}_1, \text{ and } F_2$$
$$= m_2 g + \frac{m_1}{m_2}F_2$$

Thus the Lagrangian is:

40

$$L = \frac{1}{2}m_1\left(\left(\dot{r}_1 + \ddot{r}_2\dot{\theta}^2\right)\right) + \frac{1}{2}m_2(\dot{y}_2)^2 - U_2 - U_1, \text{ where } U_2$$
$$= y_2 F_2 \text{ and } U_1 = -r_1 F_1$$

which can be rewritten:

$$L = \frac{1}{2}(m_1 + m_2)(\dot{r})^2 + \frac{1}{2}m_1 r_1^2\dot{\theta}^2 - (l - r_1)\left(\frac{m_2^2}{m_1 + m_2}\right)g$$
$$+ r_1\left(\frac{m_1 m_2}{m_1 + m_2}\right)g$$

We can drop constant terms from the Lagrangian (since they make no change in the E-L equations, thus no change in the equations of motion). So dropping the constant term and regrouping:

$$L = \frac{1}{2}(m_1 + m_2)(\dot{r})^2 + \frac{1}{2}m_1 r^2\dot{\theta}^2 + rm_2 g$$

We can now proceed with the evaluation of the Lagrangian, again starting with the conservation of angular momentum term:

$$\frac{d}{dt}\frac{\partial L}{\partial \dot{\theta}} - \frac{\partial L}{\partial \theta} = 0 \rightarrow \frac{d}{dt}(m_1 r^2\dot{\theta}) = 0 \rightarrow m_1 r^2\dot{\theta} = p_\theta$$

Thus we have:

$$L = \frac{1}{2}(m_1 + m_2)(\dot{r})^2 + \frac{p_\theta^2}{2m_1 r^2} + m_2 gr$$

The remaining equation of motion is:

$$\frac{d}{dt}\frac{\partial L}{\partial \dot{r}} - \frac{\partial L}{\partial r} = 0 \rightarrow (m_1 + m_2)\ddot{r} - m_2 g + \frac{p_\theta^2}{m_1 r^3} = 0$$

For r small we then have:

$$\ddot{r} = -\frac{p_\theta^2}{(m_1 + m_2)m_1}\frac{1}{r^3} = -\beta\frac{1}{r^3}, \quad \text{where } \beta = \frac{p_\theta^2}{(m_1 + m_2)m_1}$$

Thus, we can write:

$$\ddot{r}\dot{r} = -\beta\frac{\dot{r}}{r^3} \rightarrow (\dot{r})^2 = +\beta\left(\frac{1}{r^2}\right) \rightarrow \dot{r} = \frac{\sqrt{\beta}}{r} \rightarrow r\dot{r} = \sqrt{\beta} = \frac{1}{2}\frac{d}{dt}r^2 \rightarrow r$$
$$= \sqrt{2\sqrt{\beta}t}$$

The latter result for the r equation of motion is indicative of a repulsive potential, which then begs the question, when do we have stable orbits?

$$L = \frac{1}{2}m_1(\dot{r})^2 + \frac{p_\theta^2}{2(m_1 + m_2)r^2} + m_2 gr \rightarrow -U$$
$$= \frac{p_\theta^2}{2(m_1 + m_2)r^2} + m_2 gr,$$

Thus,

$$\frac{dU}{dr} = 0 \implies -\frac{p_\theta^2}{(m_1 + m_2)r_{eq}^3} + m_2 g = 0 \implies r_{eq} = \sqrt[3]{\gamma}, \text{ where } \gamma$$

$$= \frac{p_\theta^2}{(m_1 + m_2)m_2 g}$$

Exercise 3.13. *Could this apparatus be used to weigh unknown mass m_2? Describe a process to do this.*

Example 3.14. Revisit the single pendulum with horizontally oscillating support.
Let's now revisit the single pendulum when the point of support is oscillating horizontally. The pendulum moves in the plane of the paper. The string of length l does not bend. The point of support P moves back and forth along a horizontal direction according to the equation $x = a\cos(\omega t)$, and $(\omega \neq \sqrt{(g/l)})$:

(i) Let's start by writing the Lagrangian for this system and obtain Lagrange's equations of motion. (Don't forget the generalized force when writing Lagrange's equation for x).
Have: $x' = x + l\sin\theta$, thus $\dot{x}' = \dot{x} + l\cos\theta\dot{\theta}$. Have $y' = -l\cos\theta$, thus $\dot{y}' = l\sin\theta\,\dot{\theta} = -mgl\cos\theta$. Also have the usual $U = mgy$, to then write the Lagrangian:

$$L = \frac{1}{2}m\left(\left[-a\omega\sin(\omega t) + l\cos\theta\,\dot{\theta}\right]^2 + [l\sin\theta\dot{\theta}]^2\right)$$
$$+ mgl\cos\theta$$
$$= \frac{1}{2}ml^2\dot{\theta}^2 + mgl\cos\theta + am\omega^2 l\cos(\omega t)\sin\theta$$
$$\frac{d}{dt}\left(\frac{d}{\partial\dot{\theta}}\right) - \frac{\partial L}{\partial\theta} = 0$$
$$\rightarrow ml^2\ddot{\theta} + mgl\sin\theta$$
$$- am\omega^2 l\cos(\omega t)\cos\theta = 0$$

(ii) Next, solve the above equations of motion at first order in θ (small oscillations), and find the steady-state solution for $\theta(t)$, in terms of m, l, a, and ω. (We are not interested in the solution oscillating at the natural frequency of the pendulum.) Thus:

$$ml^2\ddot{\theta} + mgl\theta - am\omega^2 l\cos(\omega t) = 0$$
$$\ddot{\theta} + \frac{g}{l}\theta - \frac{a}{l}\omega^2\cos(\omega t) = 0.$$

So, have:

42

$$\ddot{\theta} + \frac{g}{l}\theta = \frac{a}{l}\omega^2 \cos(\omega t)$$

where the RHS is an effective force/m. And we have the solution:

$$\theta = \frac{(a/l)\omega^2}{\omega_0^2 - \omega^2}\cos(\omega t + \beta).$$

Exercise 3.14. *Repeat but with a vertically oscillating support.*

3.5 Similar Systems and the Virial theorem
So far we've seen how global symmetries play a role in establishing (additive) conservation laws. Now let's consider symmetries internal to the Lagrangian such that it can be expressed as another Lagrangian with an overall constant multiplier. In such an instance we will find that the equations of motion will be the same. To see if a Lagrangian will exhibit such a "similarity" requires a specification of the potential energy term in precisely this regard. So, let's rescale the system lengths and time, and have the potential energy be a homogeneous function of parameter rescaling (where the degree of homogeneity is given by parameter k):

$$\vec{q}_a \longrightarrow \alpha\vec{q}_a, \ (l' = \alpha l, \text{length dilation})$$
$$\dot{\vec{q}}_a \longrightarrow \left(\frac{\alpha}{\beta}\right)\dot{\vec{q}}_a, \ (t' = \beta t, \text{time dilation})$$
$$U(\alpha\{\vec{q}_a\}) \longrightarrow \alpha^k \, U(\{\vec{q}_a\}), \text{(homogeneous, degree k)}.$$
$$\text{(Eqn. 3-18abc)}$$

Now that the dilations are specified, for there to be a similarity in the Lagrangian such that an overall constant factor results, with typical Lagrangian specification $L = T - U$, we have the rescaling of the potential energy part already, the rescaling on the kinetic energy part is simply that given by the velocity above (squared). Thus, in order to have a similar system:

$$\left(\frac{\alpha}{\beta}\right)^2 = \alpha^k \longrightarrow \beta = \alpha^{1-\frac{1}{2}k}, \quad \left(\frac{E'}{E}\right) = \alpha^k \text{ and } \left(\frac{M'}{M}\right) = \alpha^{1+\frac{1}{2}k}.$$
$$\text{(Eqn. 3-19)}$$

Let's consider some cases where we have a homogeneous potential:
(1) For small oscillations, or the classic spring, the potential energy is a quadratic function of coordinates (k=2). The critical relation above with k=2 becomes: $\beta = \alpha^0 = 1$, i.e., it doesn't matter the size of the displacement from rest-position (amplitude), the time-ratio of system

will be 1, i.e. the system period is independent of amplitude.

(2) For a uniform field of force the potential energy is a linear function of coordinates, such as the approximation for motion due to gravity near the Earth's surface (P.E. = mgh). For k=1 we have: $= \sqrt{\alpha}$, so in-fall under gravity. Time of fall, for example, goes as the square root of the initial height.

(3) For the Newtonian or Coulomb potential: k = -1. Now have $= \sqrt[3]{\alpha}$, the square of the period of an orbit goes as the cube of the size of the orbit (Kepler's 3rd Law).

Virial theorem

This is the one of the few examples, or contexts, where a multi-element system is being considered (and for very large numbers of elements), due to its universal application. Any homogeneous potential where the motion is bounded allows for the Virial Theorem to apply, whereby time averages of the potential and kinetic energy of the system have a simple relation. This will be derived as follows, consider:

$$E = \sum_i \left(\dot{q}_i \frac{\partial L}{\partial \dot{q}_i} \right) - L \implies \sum_i \left(\dot{q}_i \frac{\partial L}{\partial \dot{q}_i} \right) = 2T$$

(Eqn. 3-20)

Writing $v_i = \dot{q}_i$ and the definition of generalized momenta, then switching to vector notation with particles indicated by indexing 'a':

$$\sum_i (v_i \, p_i) = \sum_a \vec{v}_a \cdot \vec{p}_a = \frac{d}{dt} \left(\sum_a \vec{r}_a \cdot \vec{p}_a \right) - \sum_a \vec{r}_a \cdot \dot{\vec{p}}_a$$

Let's now take the time average of 2T, where the total time derivative term will have mean value zero if we have bounded motion. To be specific, the time average for a function $f(t)$ of time is defined to be:

$$\overline{f} = \lim_{\tau \to \infty} \frac{1}{\tau} \int_0^\tau f(t) dt$$

(Eqn. 3-21)

Suppose $f(t) = \frac{d}{dt} F(t)$, then:

$$\overline{f} = \lim_{\tau \to \infty} \frac{1}{\tau} [F(\tau) - F(0)] = 0 \quad (for \; bounded \; motion).$$

Since we have bounded motion if we stay in a finite region of space with finite velocities, we then have:

44

$$2\overline{T} = -\sum_a \vec{r}_a \cdot \dot{\vec{p}}_a = \sum_a \vec{r}_a \cdot \frac{\partial U}{\partial \vec{r}_a} = k\overline{U}$$

Revisiting what this indicates for the three cases mentioned above ($E = \overline{E} = \overline{T} + \overline{U}$):

(1) Small oscillations (k=2), have $\overline{T} = \overline{U}, E = 2\overline{T}$.

(2) Uniform field (k=1), have $\overline{T} = (1/2)\overline{U}, E = 3\overline{T}$

(3) Newtonian or Coulomb potential (k = −1): $\overline{U} = -2\overline{T}, E = -\overline{T}$. This result is consistent with the total energy of a bounded motion in this type of potential being negative, as will be apparent in the examples that follow.

3.6. One-dimensional systems

Often system analysis reduces in dimensionality (due to symmetries). Consider the orbit of a planet about the sun, where the 3D problem reduces to 2D problem by conservation of angular momentum. For the most part, we only need consider motion in one or two dimensions. Let's start with one-dimensional motion.

Consider the following Lagrangian for one dimensional motion where an arbitrary potential is sketched as shown in Fig. 3.5.

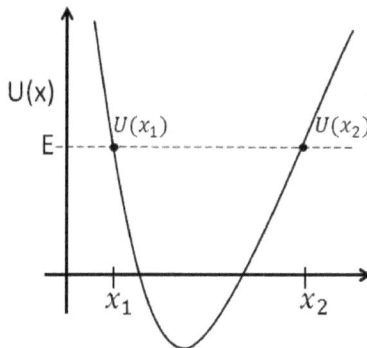

Fig. 3.5. A One-dimensional potential. $U(x_1) = E = U(x_2)$.

$$L = \frac{1}{2}m\,\dot{x}^2 - U(x) \longrightarrow E = \frac{1}{2}m\,\dot{x}^2 + U(x)$$

(Eqn. 3-22)

Since $U(x) \leq E$, and taking the positive root (the negative corresponds with time-reversal, with same type of solutions):

45

$$\frac{dx}{dt} = \sqrt{\frac{2[E - U(x)]}{m}} \rightarrow t = \sqrt{m/2} \int dx/\sqrt{E - U(x)} + C$$

The limits of motion are given by $U(x_1) = E = U(x_2)$, and the period of motion is given by twice the integral from x_1 to x_2:

$$Period = \sqrt{2m} \int_{x_1}^{x_2} dx/\sqrt{E - U(x)}.$$

(Eqn. 3-23)

Example 3.15. Motion on a curved ramp.

A small mass slides without friction on a block of mass M as shown in Fig. 3.6. M itself slides without friction on a horizontal table, and its curved side has the shape of a circle of radius a.

    a) Find Lagrange's equations for the system in terms of two generalized coordinates.

    b) Find two conserved quantities.

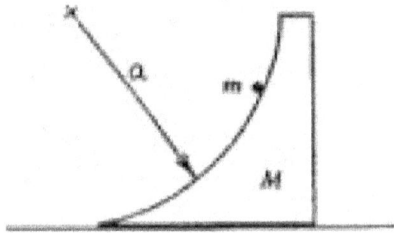

Fig. 3.6. A mass m slides without friction on a block of mass M, with circle of radius a.

The coordinates: $x_1 = x + a \cos \theta$; $y_1 = -a \sin \theta$; and $x_2 = x$.
The coordinate time derivatives: $\dot{x}_1 = \dot{x} + a \sin \theta \, \dot{\theta}$; $\dot{y}_1 = -a \cos \theta \, \dot{\theta}$; and $\dot{x}_2 = \dot{x}$.
The potential energy: $U = -mga \sin \theta$.
Thus,

$$L = T - U = \frac{1}{2}m \left([\dot{x} - a \sin \theta \, \dot{\theta}]^2 + [-a \cos \theta \, \dot{\theta}]^2 \right) + \frac{1}{2}M(\dot{x})^2 - U$$

$$L = \frac{1}{2}(m + M)\dot{x}^2 + \frac{1}{2}m(a\dot{\theta})^2 - am\dot{x}\dot{\theta} \sin \theta + mga \sin \theta$$

and,

$$\frac{d}{dt}\left(\frac{\partial L}{\partial \dot{x}}\right) - \frac{\partial L}{\partial x} = 0 \Rightarrow (m + M)\ddot{x} - \frac{d}{dt}(am\dot{\theta} \sin \theta) = 0, \text{ thus,}$$

$$\frac{d}{dt}\{(m + M)\dot{x} - am\dot{\theta} \sin \theta\} = 0.$$

46

So, we have:
$$(m + M)\dot{x} - am\dot{\theta}\sin\theta = const,$$
and,
$$E = T + U = \frac{1}{2}(m + M)\dot{x}^2 + \frac{1}{2}m(a\dot{\theta})^2 - am\dot{x}\dot{\theta}\sin\theta + mga\sin\theta.$$

Exercise 3.15. *Find the velocities of the masses as a function of time when mass m released from rest at the top of the curved side.*

3.7 Motion in a Central Field

Consider a single particle in a central potential. Its angular momentum is conserved: $\vec{M} = \vec{r} \times \vec{p} = constant$. Since constant $\vec{M}$ is perpendicular to $\vec{r}$, the position is always in a plane perpendicular to $\vec{M}$ (conservation of angular momentum has thereby reduced the problem from 3D to 2D). The appropriate form for the Lagrangian for motion in a plane with central potential is thus:

$$L = \frac{1}{2}m\dot{r}^2 + \frac{1}{2}m(r\dot{\varphi})^2 - U(r)$$

(Eqn. 3-24)

Notice that there is no direct reference to the coordinate φ, in the Hamiltonian formalism this means that:

$$F_\varphi = \frac{\partial L}{\partial \varphi} = 0$$

thus

$$\dot{p}_\varphi = F_\varphi = 0 \rightarrow p_\varphi = constant = \text{"}M\text{"}.$$

$$p_\varphi = \frac{\partial L}{\partial \dot{q}_i} = mr^2\dot{\varphi} = M.$$

(Eqn. 3-25)

Recall that the area of a radial sector radius r with angle of sweep φ is $A = (1/2)r \cdot r\varphi$, and the sectorial velocity is thus $V_{sectorial} = (1/2)r^2\dot{\varphi} = M/2m$ is a constant, i.e., "equal areas swept in equal times", a.k.a. Kepler's Third Law. As is typical in this type of analysis, integrals of the motion (e.g., conservation laws) are used as a first step to simplify analysis. Thus, for energy we have:

$$E = \frac{1}{2}m\dot{r}^2 + \frac{1}{2}m(r\dot{\varphi})^2 + U(r) \quad \rightarrow \quad \frac{1}{2}m\dot{r}^2 = [E - U] - \frac{M^2}{2mr^2},$$

where the last term is the centrifugal energy. Rearranging:

$$\frac{dr}{dt} = \sqrt{\frac{2}{m}[E - U] - \frac{M^2}{m^2 r^2}}$$

Integrating, we get

$$t = \int \frac{dr}{\sqrt{\frac{2}{m}[E - U] - \frac{M^2}{m^2 r^2}}} + C_1$$

(Eqn. 3-26)

Using $d\varphi = \frac{M}{mr^2} dt$,

$$\varphi = \int \frac{M dr / r^2}{\sqrt{2m[E - U] - \frac{M^2}{r^2}}} + C_2$$

(Eqn. 3-27)

Note, $\dot{\varphi} = M$ means φ changes monotonically, so for a closed path, that necessarily has a (bounded) minimum and maximum radius, we have for change in phase in going from the minimum radius to the maximum radius and then back:

$$\Delta\varphi = 2 \int_{r_{min}}^{r_{max}} \frac{M dr / r^2}{\sqrt{2m[E - U] - \frac{M^2}{r^2}}}$$

where the limits of the motion are given by the energy having no kinetic part, $E = U_{eff}$ where

$$U_{eff} = U + \frac{M^2}{2mr^2}.$$

(Eqn. 3-28)

The $\Delta\varphi$ for there to result in a closed path must be exactly equal to 2π or a multiple of $\Delta\varphi$ must result in a multiple of 2π (i.e. $\Delta\varphi = 2\pi (m/n)$). This only happens for all paths in the integral above when the potentials U have the form $1/r$ or r^2, and in those instances an extra integral of the motion occurs (known as the Runge-Lens vector). Before moving on to the critical $1/r$ potential, however, let's consider the implications of nonzero angular momentum with a central potential. It is generally impossible to reach the center in such instances, even in attractive potentials. To reach the center when $M \neq 0$, we are obviously considering a situation where we are not at the turning points of the motion, thus

$$\frac{1}{2}m\dot{r}^2 = [E - U] - \frac{M^2}{2mr^2} > 0,$$

and regrouping and taking the limit as radius goes to zero, we find that the only potentials allowing this must satisfy:

$$\lim_{r \to 0} r^2 U < -\frac{M^2}{2m}$$

This is only possible for negative potentials $U(r) = -\alpha/r^n$ with $n > 2$ or with $n = 2$ and $\alpha > \frac{M^2}{2m}$.

In the preceding example we saw that the Kepler and Coulomb potentials $(U(r) = -\alpha/r)$ were not in the group of potentials allowing motion through the center when angular momentum is nonzero. Let's now consider the attractive potential relevant for gravity (and for attraction between opposite charges) with $U(r) = -\alpha/r$ in more detail. To start, the angle integral can be easily solved for this situation, where the effective potential is:

$$U_{eff} = -\frac{\alpha}{r} + \frac{M^2}{2mr^2}, \text{ and } \min_r U_{eff} = -\frac{m\alpha^2}{2M^2} \text{ at } r = \frac{M^2}{m\alpha}$$

$$\text{(Eqn. 3-29)}$$

where the function minimum and significant energy domains are indicated in Fig. 3.7.

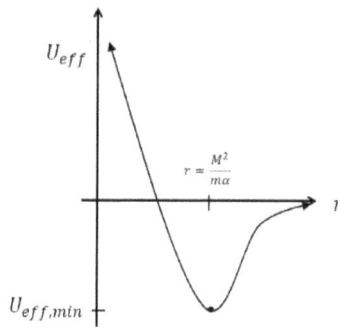

Fig. 3.7. A sketch of the effective potential. $U_{eff,min} = -\frac{m\alpha^2}{2M^2}$. The motion is finite if $E < 0$, infinite if $E \geq 0$.

49

Integration then yields:

$$\varphi = \cos^{-1} \frac{\left(\frac{M}{r} - \frac{m\alpha}{M}\right)}{\sqrt{2mE + \frac{m^2\alpha^2}{M^2}}} + constant$$

(Eqn. 3-30)

Let's have $\varphi = 0$ correspond to occurrence of nearest approach (perihelion, r_{min} in what follows), in which case the constant is zero. Let's also relate to two forms of describing orbits $\{p, e\}$, where $2p$ is known as the latus rectum, and e is the eccentricity, and the conic section parameters $\{a, b\}$, where $2a$ is the length of the major axis and $2b$ is the length of the minor axis:

$$p = \frac{M^2}{m\alpha} \quad and \quad e = \sqrt{1 + \frac{2EM^2}{m\alpha^2}}$$

(Eqn. 3-31)

to arrive at the orbit equation:

$$p = r(1 + e \cos \varphi)$$

(Eqn. 3-32)

From the orbit equation we can see that:

$$r_{min} = \frac{p}{1 + e} \quad and \quad r_{max} = \frac{p}{1 - e}$$

(Eqn. 3-33)

Since $2a = r_{min} + r_{max}$:

$$a = \frac{p}{1 - e^2} = \frac{\alpha}{2|E|}$$

(Eqn. 3-34)

We also see that the ratios b/r_{min} and r_{max}/b are rescale invariants and must be proportional to each other, where for $e = 0$ this is shown to be equality, thus $b = \sqrt{r_{min} \cdot r_{max}}$ and we get:

$$b = \frac{p}{\sqrt{1 - e^2}} = \frac{M}{\sqrt{2m|E|}}$$

(Eqn. 3-35)

Let's now consider the various cases in terms of the eccentricity parameter $e = \sqrt{1 + \frac{2EM^2}{m\alpha^2}}$ of the orbit:

For $e = 0$ (occurs when $E = -\frac{m\alpha^2}{2M^2}$): We have a circular orbit $r_{min} = r_{max} = p$.

50

For $0 \leq e < 1$ (occurs when $E < 0$): We have elliptical orbit $r_{min} \neq r_{max}$.

For ellipses and the circle we have bound orbits, which allows us to do the full sectorial integral of one such orbit, thereby getting simply the area of the ellipse or circle. Recall

$$\frac{d(area)}{dt} = V_{sectorial} = \frac{1}{2}r^2\dot{\varphi} = \frac{M}{2m}$$

(Eqn. 3-36)

integrating over the time of one orbital period T:

$$T = \frac{2m(area)}{M} = \frac{2m\pi ab}{M} = \pi a\sqrt{\frac{m}{2|E|^3}}.$$

(Eqn. 3-37)

From this exact solution we can see that $T^2 \propto \frac{1}{|E|^3} \propto a^3$, which is Kepler's 3rd Law.

For $e = 1$ (occurs when $E = 0$): We have a parabolic orbit (unbounded) with $r_{min} = \frac{p}{2}$ and $r_{max} = \infty$, which describes an infalling particle from rest at infinity.

For $e > 1$ (occurs when $E > 0$): We have a hyperbolic orbit (unbounded).

The Laplace-Runge-Lenz Vector

Consider an inverse-square central force acting on a single particle that is described by the equation

$$A = p \times L - mk\hat{r} \rightarrow e = \frac{A}{mk},$$

(Eqn. 3-38)

where

m is the mass of the point particle moving under the central force,
p is its momentum vector,
$L = r \times p$ is its angular momentum vector,
r is the position vector of the particle (Fig. 3.8),
$\hat{r}$ is the corresponding unit vector, i.e., $\hat{r}$, and
r is the magnitude of r, the distance of the mass from the center of force.

51

The constant parameter k describes the strength of the central force; it is equal to $\underline{G} \cdot M \cdot m$ for gravitational and $-\underline{k_e} \cdot Q \cdot q$ for electrostatic forces. The force is attractive if $k > 0$ and repulsive if $k < 0$.

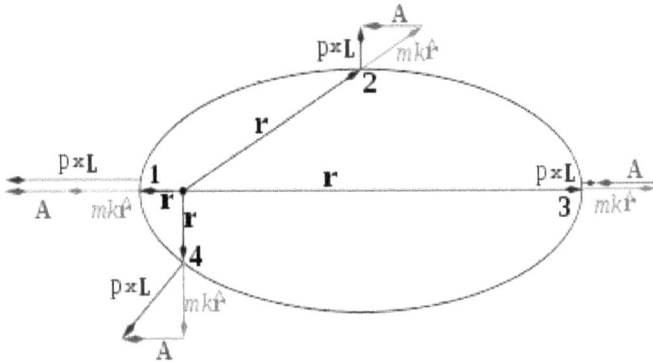

Fig. 3.8. The LRL vector **A** at four points on the elliptical orbit under an inverse-square central force. The center of attraction is shown as a small black circle from which the position vectors emanate. The angular momentum vector **L** is perpendicular to the orbit. The coplanar vectors $\mathbf{p} \times \mathbf{L}$ and $(mk/r)\mathbf{r}$ are shown. The vector **A** is constant in direction and magnitude.

The seven scalar quantities E, **A** and **L** (being vectors, the latter two contribute three conserved quantities each) are related by two equations, $\mathbf{A} \cdot \mathbf{L} = 0$ and $A^2 = m^2k^2 + 2\,mEL^2$, giving five independent <u>constants of motion</u>. This is consistent with the six initial conditions (the particle's initial position and velocity vectors, each with three components) that specify the orbit of the particle, since the initial time is not determined by a constant of motion. The resulting 1-dimensional orbit in 6-dimensional phase space is thus completely specified.

Example 3.16. A test mass is released over the north pole.
A test mass is released at rest, one earth diameter above the (rotational) north pole. Ignore atmospheric friction. (Use for acceleration of gravity near the earth's surface $10 \dfrac{m}{sec^2}$, and for radius of the earth $R_e = $ 6,400 km.)
  a) Find the velocity (in meter/sec) of the mass when it hits the earth.

b) Find an expression for the length of time that it takes the mass to hit the earth. Your expression should contain a dimensionless integral.

Solution:
(a) Velocity at surface of earth: the potential energy of the test mass: $\Phi = -\frac{mGM}{R}$. Conservation of energy gives the kinetic energy to be the change in potential energy:

$$\frac{1}{2}mv^2 = \Delta PE = (\frac{-mGM}{R})\Big|_{R_e}^{3R_e} = \frac{2}{3}m\,R_e\,g$$

(b) Time until impact, let's first get the relation for drop to radius r:

$$\frac{1}{2}mv^2 = (\frac{-mGM}{R})\Big|_r^{3R_e} \qquad v$$

$$= \frac{dr}{dt}\; since\; no\; coriolis\; force\; at\; North\; pole$$

$$\frac{1}{2}m\left(\frac{dr}{dt}\right)^2 = \frac{mGM}{r} - \frac{mGM}{3R_e}$$

$$\frac{dr}{dt} = \sqrt{\frac{2GM}{r} - \frac{2GM}{3R_e}} = \sqrt{2GM}\sqrt{\frac{1}{r} - \frac{1}{3R_e}}$$

$$dt = \frac{1}{\sqrt{2GM}}\frac{dr}{\sqrt{\frac{1}{r} - \frac{1}{3R_e}}}$$

$$T = \frac{1}{\sqrt{2GM}}\int_{R_e}^{3R_e}\frac{dr}{\sqrt{\frac{1}{r} - \frac{1}{3R_e}}} = \frac{(3R_e)^{\frac{3}{2}}}{\sqrt{2GM}}\int_{\frac{1}{3}}^{1}\frac{dx}{\sqrt{\frac{1}{x} - 1}} \cong 1.43\frac{(3R_e)^{\frac{3}{2}}}{\sqrt{2GM}}$$

Exercise 3.16. *A test mass is released over the equator.*

Example 3.17. A planet of mass M....
A planet of mass m orbits a sun mass M. We saw in the general properties of Keplerian systems that the planet moves in a plane containing the center of force. (a) Introduce polar coordinates for the plane of motion and write the Lagrangian; (b) Obtain the angular momentum and energy of the planetary system; and (c) from the Keplerian analysis we know that the orbit is an ellipse, so relate the semimajor axis length a and eccentricity ε of that ellipse to the conserved energy and angular moment obtained in (b), using the following parameterization of the orbit as an ellipse:

53

$$\frac{1}{e} = \frac{1}{a(1-\varepsilon^2)} + \frac{\varepsilon}{a(1-\varepsilon^2)}\cos\theta$$

Solution:

(a) We have from Newtonian gravitational force and we shift to the center of mass frame:

$$F = \frac{mMG}{r^2} = \frac{M_T\mu G}{r^2}, \text{where } M_T = (m+M) \text{ and } \mu = \frac{mM}{m+M}$$

For this we can write the potential energy as:

$$U = -\frac{M_T\mu G}{r}$$

So, in polar coordinates the Lagrangian $L = T - U$:

$$L = \frac{1}{2}\mu\left(\dot{r}^2 + r^2\dot{\theta}^2\right) - U(|\vec{r}|) \text{ and } \vec{r} = \vec{r}_m - \vec{r}_M, r = |\vec{r}|$$

(b) To get the energy let's start with obtaining the equations of motion for the cyclic coordinates, here the orbital angle, to get other constants of the motion, then use $E = T + U$:

$$\frac{d}{dt}\left(\mu r^2\dot{\theta}\right) = 0 \rightarrow l = \mu r^2\dot{\theta}, angular\ momemtum\ conserved$$

$$E = \frac{1}{2}\mu\dot{r}^2 + \frac{l^2}{2\mu r^2} - \frac{\mu M_T G}{r}$$

(c) Relation to parameterization of an ellipse. At r_{min} and r_{max} we have $\dot{r} = 0$, so get:

$$E = \frac{l^2}{2\mu r_{min}^2} - \frac{\mu M_T G}{r_{min}} \text{ and } E = \frac{l^2}{2\mu r_{max}^2} - \frac{\mu M_T G}{r_{max}}$$

From the ellipse parameterization we have for r_{min} and r_{max}:

$$\frac{1}{r_{min}} = \frac{1}{a(1-\varepsilon^2)} + \frac{\varepsilon}{a(1-\varepsilon^2)} \implies r_{min} = a(1-\varepsilon)$$

$$\frac{1}{r_{max}} = \frac{1}{a(1-\varepsilon^2)} + \frac{\varepsilon}{a(1-\varepsilon^2)} \implies r_{max} = a(1+\varepsilon)$$

Using with the two equations for energy at the max and min r positions we get:

$$\frac{l^2}{2\mu}\left(\frac{1}{r_{max}{}^2} - \frac{1}{r_{max}{}^2}\right) - \mu M_T G\left(\frac{1}{r_{min}} - \frac{1}{r_{max}}\right) = 0 \quad \rightarrow \quad l^2$$
$$= \mu^2 M_T Ga(1 - \varepsilon^2)$$

Substituting the relation for l^2 into the two energy equations, as well as $r_{min} = a(1 - \varepsilon)$ and $r_{max} = a(1 + \varepsilon)$, we get:
$$E = \frac{-\mu M_T G}{r_{min} + r_{max}} = \frac{-\mu M_T G}{2a}$$

Thus,
$$a = \frac{-\mu M_T G}{2E} = \frac{mMG}{2|E|} = \frac{\alpha}{2|E|}, where\ a = \mu M_T G = mMG.$$

And substituting into the l^2 relation we regroup to the get expressin for eccentricity:
$$\varepsilon = \sqrt{1 + \left(\frac{2El^2}{\mu\alpha^2}\right)}.$$

Exercise 3.17. *What is the eccentricity of the Earth-Moon system? Of the Earth-Sun system?*

Example 3.18. A particle of mass m...
A particle of mass m moves in a potential $U = \alpha/r - \beta/r^3$, $\alpha, \beta > 0$.
   a) For what range a radii, r, are circular orbits stable? (Express the condition on r in terms of α and β.)
   b) Find in terms of r, α, β, and m the frequency Ω of a circular orbit and the frequency w of small oscillations about a circular orbit.

Solution:
(a) $U = \alpha/r - \beta/r^3$, $\alpha, \beta > 0$, and for orbits: $L = \frac{1}{2}m(\dot{r}^2 + r^2\dot{\theta}^2) - U$
and $E = \frac{1}{2}m\dot{r}^2 + \frac{M_\theta^2}{2mr^2} + U$, thus
$$U_{eff} = \frac{M_\theta^2}{2mr^2} - \frac{\alpha}{r} - \frac{\beta}{r^3}.$$
Circular orbits for:
$$\frac{\partial U_{eff}}{\partial r} = 0 \quad \rightarrow \quad -\frac{M_\theta^2}{mr^3} + \frac{\alpha}{r^2} + \frac{3\beta}{r^4} = 0$$
Stable orbits for:

$$\frac{\partial^2 U_{eff}}{\partial r^2} = \frac{3M_\theta^2}{mr^4} - \frac{2\alpha}{r^3} - \frac{12\beta}{r^5} > 0.$$

(b) Recall the area swept, A, relation: $M_\theta = mr^2\dot\theta = 2m\frac{dA}{dt}$, then can write:

$$dt = \frac{2m}{M_\theta}dA \Rightarrow T = \frac{2m}{M_\theta}(\pi r_c^2)$$

$$\alpha r_c^2 - \frac{M_\theta^2}{m}r_c + 3\beta = 0$$

The frequency of the circular orbit, Ω, is:

$$\Omega = \frac{2\pi}{T} = \frac{M_\theta}{mr_c^2},$$

and the frequency of small oscillations about that circular orbit:

$$\omega = \sqrt{\frac{1}{2m}\frac{\partial^2 U_{eff}}{\partial r^2}\bigg|_{r_c}} = \sqrt{\frac{1}{m}\left\{\frac{\alpha}{r^3} - \frac{3\beta}{r^5}\right\}}.$$

Exercise 3.18. *What happens when* α *and* β *.are selected such that* $\Omega = \omega$?

Example 3.19. Particle in a central force field.
A particle moves in a central force field given by the potential: $U = -K\frac{e^{-r/a}}{r}$, where K and a are positive constants. (a) Find the relation between r, l, and E for circular orbits. (b) Find the period of small oscillations (in the r-θ plane) about a circular orbit.

Solution:

(a) So, have $U = -K\frac{e^{-r/a}}{r}$ and $L = \frac{1}{2}m(\dot r^2 + r^2\dot\theta^2) - U$. For the centrifugal barrier we have:

$$\frac{d}{dt}\left(\frac{\partial L}{\partial \dot\theta}\right) = 0 \Rightarrow mr^2\dot\theta = |L|$$

So,

$$L = \frac{1}{2}m\dot r^2 - \frac{|L|^2}{2mr^2} - U$$

and the equations of motion are:

$$\frac{d}{dt}(m\dot r) - \left\{-\frac{|L|^2}{mr^3} - \frac{\partial U}{\partial r}\right\} = 0$$

Have circular orbits $r = const$ for:

$$\frac{|L|^2}{mr_0^3} = -\frac{\partial U}{\partial r}\Big|_{r=r_0} \;\rightarrow\; \frac{l^2}{mr_0^3} + \frac{E}{r_0} = +\frac{K}{ar_0}e^{-r_0/a} \;\rightarrow\; E$$

$$= \frac{l^2}{2mr_0^2} + \frac{K}{a}e^{-r_0/a}$$

(b) We have $\omega = \sqrt{\dfrac{1}{2m}\dfrac{\partial^2 U_{eff}}{\partial r^2}}$ and $U_{eff} = \dfrac{+l^2}{2mr^2} - \dfrac{Ke^{-r/a}}{r}$, and at oscillation equilibrium:

$$\frac{U_{eff}}{\partial r} = \frac{-l^2}{mr^3} + \frac{Ke^{-r/a}}{r^2} + \frac{Ke^{-r/a}}{ar} = 0,$$

thus,

$$\frac{\partial^2 U_{eff}}{\partial r^2} = \frac{3l^2}{mr^4} - \frac{2Ke^{-r/a}}{r^3} - \frac{Ke^{-r/a}}{ar^2} - \frac{Ke^{-r/a}}{ar^2} - \frac{Ke^{-r/a}}{a^2r}.$$

From

$$\left(\frac{1}{r^2} + \frac{1}{ar}\right)Ke^{-r/a} = \frac{l^2}{mr^3} \quad and \quad Ke^{-r/a} = \left(\frac{ar}{a+r}\right)\frac{l^2}{mr^2}$$

$$= \frac{a}{a+r}\frac{l^2}{mr}$$

We can then regroup to get

$$\omega = \sqrt{\frac{l^2}{m^2r^2}\left\{\frac{a}{a+r}\right\}\left(\frac{1}{r^2} + \frac{1}{ar} - \frac{1}{a^2r}\right)}.$$

Exercise 3.19. *Suppose* $\dfrac{\partial^2 U_{eff}}{\partial r^2}\Big|_{r_c}$ *for some choice of K and a, derive frequency formula to third order derivative in potential, what is the new oscillatory frequency?*

Example 3.20. Kepler's 3rd Law from Newton's laws.
(a) Show directly from Newton's laws that, for two stars of mass m1 and m2 in circular orbits about their center of mass, Kepler's 3rd Law has the form: $T^2 = \dfrac{4\pi^2}{G(m_1+m_2)}R^3$, with T the period and R the distance between the stars.
(b) Show that the formula can be rewritten in the form $T^2 = (m_1 + m_2)^{-1}R^3$, with T in years, R in AU (astronomical units), and m in solar masses. (If R is the semimajor axis, this holds for elliptical orbits as well.)

(c) Show that for a small object in circular orbit at the surface of a large object, $T = K\rho^{-1/2}$, and find the constant K. What is the period of a pebble in orbit at the surface of a spherical rock ($\rho = 3g/cm^3$)?

Solution:

(a) Recall: $L = r \times \mu v = \text{const}$ and $dA = \frac{1}{2} r \cdot r d\theta$

So,

$$L = \mu r \times (\dot{r}\hat{r} + r\dot{\theta}\hat{\theta}) = \mu r^2 \dot{\theta} = 2\mu \frac{dA}{dt} = \text{const}$$

$$2\mu dA = L dt \rightarrow 2\mu(\pi ab) = LT$$

Recall the relation of masses to the major and minor axes:

$$a = \frac{G(m_1 + m_2)\mu}{2|E|} \qquad b = \frac{L}{\sqrt{2\mu|E|}}$$

Thus,

$$LT = 2\mu\pi \frac{G(m_1 + m_2)\mu}{2|E|} \frac{L}{\sqrt{2\mu|E|}}$$

$$\rightarrow \frac{4\pi^2}{G(m_1 + m_2)} \left\{ \frac{G(m_1 + m_2)\mu}{2|E|} \right\}^3 = T^2$$

Thus, substituting for a = R (evaluation on semimajor axis):

$$T^2 = \frac{4\pi^2}{G(m_1 + m_2)} R^3.$$

(b) Unit Change goes as follows:

$$T^2 \left(\frac{365 \times 24 \times 3600\text{sec}}{1yr} \right)^2$$

$$= \frac{4\pi^2}{G(m_1 + m_2)\left(\frac{2 \times 10^{30}kg}{M_\odot}\right)} R^3 \left(\frac{1.5 \times 10^8 km}{1A.U.} \right)^3,$$

so, $T^2 = (m_1 + m_2)^{-1} R^3 K$ and $K = $

$$\frac{(1.5\times10^8 km)^3 4\pi^2}{6.67\times10^{-11} Nm^2/kg^2 (3.15\times10^7 sec)^2 (2\times10^{30} kg)} \left[\frac{M_\odot \cdot yr^2}{(A.U.)^3}\right] = 1.0 \left[\frac{M_\odot \cdot yr^2}{(A.U.)^3}\right].$$

Thus,

$$T^2 = (m_1 + m_2)^{-1} R^3.$$

(c) $T^2 = (m_1 + m_2)^{-1} R^3 \simeq m_{Large}^{-1} R^3 \simeq \dfrac{\frac{4}{3}\pi R^3}{m_{Large}} \dfrac{1}{\frac{4}{3}\pi} = \dfrac{\rho}{\frac{4}{3}\pi}$, thus $T =$

$K\rho^{-1/2}$ where $K = \dfrac{1}{2\sqrt{\frac{\pi}{3}}}$ (where T is in units of years, $R = AU's$, $m =$

$M_\odot's$, and $m_1 \gg m_2$. For $\rho = 3g/cm^3 = 3 \times 10^3 kg/m^3$, thus:

$$T = \sqrt{\dfrac{3\pi}{6.67 \times 10^{-11}}} (3 \times 10^3)^{-1/2} sec = 6.86 \times 10^3 sec = 114 \ min.$$

Exercise 3.20. What is the period of a pebble in orbit at the surface of earth ($\rho = 1g/cm^3$) and at the surface of a neutron star ($\rho = 10^{16} g/cm^3$)?

Example 3.21. Binary systems.
Stellar masses are found by observing binary systems. Typically one cannot resolve the stars, but the spectrum shows two periodically changing Doppler shifts, giving the line-of-sight velocity of each star. Call the velocities V_1 and V_2. Show that if the orbit is inclined by an angle θ to the line of sight:
$$R = (V_1 + V_2)/\Omega \sin \theta \quad \text{and} \quad M_2/M_1 = V_1/V_2 \quad \text{and}$$
$$\dfrac{m_2^3}{(m_1+m_2)^2} \sin^3 \theta = (a_1 \sin \theta)^3/T^2.$$

Start with: $V_1 = \mho_1 \sin \theta$ and $V_2 = \mho_2 \sin \theta$, where $\mho_1 = r_1 \Omega$ and $\mho_2 = r_2 \Omega$. Let $R = r_1 + r_2$, then:

$$V_1 + V_2 = (\mho_1 + \mho_2) \sin \theta = R\Omega \sin \theta \rightarrow R = (V_1 + V_2)/\Omega \sin \theta$$

With the origin at CM: $M_1 r_1 + M_2 r_2 = 0$ and $M_1 \mho_1 + M_2 \mho_2 = 0$, thus: $|M_1 V_1/\sin \theta| = |M_2 V_2/\sin \theta|$
and $\dfrac{M_2}{M_1} = \dfrac{V_1}{V_2}$. To get the last relation, recall that on the semimajor axis (for R):
$$T^2 = (m_1 + m_2)^{-1} R^3,$$
thus:

$$T^2 = (m_1 + m_2)^{-1} \left\{ \dfrac{(V_1 + V_2)}{\Omega \sin \theta} \right\}^3 = (m_1 + m_2)^{-1} \left\{ \dfrac{\left(1 + \frac{m_1}{m_2}\right) V_1}{\Omega \sin \theta} \right\}^3$$

$$= (m_1 + m_2)^{-1} \left(1 + \dfrac{m_1}{m_2}\right)^3 a_1^3$$

59

From which we get:

$$\frac{m_2^3}{(m_1 + m_2)^2} \sin^3 \theta = \frac{(a_1 \sin \theta)^3}{T^2}.$$

Exercise 3.21. *Binary with neutron star.*
Consider a binary with one neutron star. The observed Doppler shift of
the neutron star has magnitude $\frac{\Delta \lambda}{\lambda} = 2 \times 10^{-6}$ and a period of 4 days. If
the mass of the neutron star is less than $3M_\Theta$, what is the maximum mass
of its companion?

Example 3.22. Motion inside a paraboloid of revolution.
A particle of mass m is constrained to move under gravity without friction
on the inside of a paraboloid of revolution whose axis is vertical. Find the
one-dimensional problem equivalent to its motion. What is the condition
on the particles initial velocity to produce circular motion? Find the
period of small oscillations about this circular motion.

Let's adopt cylindrical coordinates: $x = \rho \sin \theta$, $y = \rho \cos \theta$, in which
case we have coordinates:
$z = \frac{a}{2}\rho^2$, $\rho^2 = x^2 + y^2$, $y = x^2$, and potential $U = mgz$. Thus, the
Lagrangian is:

$$L = \frac{1}{2}m(\dot{x}^2 + \dot{y}^2 + \dot{z}^2) - mg\frac{a}{2}\rho^2,$$

where

$$\dot{z} = a\rho\dot{\rho}, \quad \dot{x} = \dot{\rho}\sin\theta + \rho\cos\theta\,\dot{\theta}, \quad \dot{y} = \dot{\rho}\cos\theta + \rho\sin\theta\,\dot{\theta}.$$

Thus,

$$L = \frac{1}{2}m\left(\dot{\rho}^2 + (a\rho\dot{\rho})^2 + \left(\rho\dot{\theta}\right)^2\right) - mg\frac{a}{2}\rho^2$$

Using the Euler-Lagrange equation for θ:

$$\frac{d}{dt}\left(\frac{\partial L}{\partial \dot{\theta}}\right) - \frac{\partial L}{\partial \theta} = 0 \quad gives \quad m\rho^2\dot{\theta} = M_\theta.$$

Thus,

$$L = \frac{1}{2}m(\dot{\rho}^2 + (a\rho\dot{\rho})^2) + \frac{1}{2}m\left(\rho\dot{\theta}\right)^2 - mg\frac{a}{2}\rho^2$$

Using the Euler-Lagrange equation for ρ we get:

$$m\ddot{\rho} + \frac{d}{dt}(m(a\rho)^2\dot{\rho}) - m(a\dot{\rho})^2\rho - m\rho\dot{\theta}^2 + mga\rho = 0$$

60

$$m\ddot{\rho}(1 + a^2\rho^2) + ma^2\rho\dot{\rho}^2 - \frac{M_\theta^2}{m\rho^3} + mga\rho = 0$$

Circular motion $\dot{\rho} = 0$:

$$\left(\frac{M_\theta}{m\rho}\right)^2 = ga\rho^2 \quad \text{and} \quad M_o = m\rho v.$$

Thus

$$v = \rho\sqrt{ga} = \sqrt{2gz}$$

Let's now consider small oscillations for

$$m\ddot{\rho}(1 + a^2\rho^2) + ma^2\rho\dot{\rho}^2 - \frac{M_\theta^2}{m\rho^3} + mga\rho = 0$$

Let $\rho = \rho_o + \eta$, then retaining terms to 1st order in η:

$$(1 + a^2\rho_o^2)m\ddot{\eta} - \frac{M_\theta^2}{m\rho_o^3}\left(1 - \frac{3\eta}{\rho_o}\right) + mga(\rho_o + \eta) = 0$$

Thus,

$$\ddot{\eta} + \frac{4ga\eta}{(1 + a^2\rho_o^2)} = 0 \quad \Longrightarrow \quad \omega = \sqrt{\frac{4ga}{(1 + a^2\rho_o^2)}} \quad \Longrightarrow \quad T$$

$$= \pi\sqrt{\frac{(1 + a^2\rho_o^2)}{ga}}.$$

Exercise 3.22. In-fall time.
Two particles move about each other in circular orbits under the influence of gravitational forces, with a period T. Their motion is suddenly stopped and they are released and allowed to fall into each other. Show they collide in time $t/4\sqrt{2}$.

Example 3.23. Attractive central force.
(a) Show that if a particle describes a circular orbit under the influence of an attractive central force directed to a point on the circle, then the force varies as the inverse fifth power of the distance.
(b) Show that for the orbit described the total energy of the particle is zero.
(c) Find the period of the motion.
(d) Find $\dot{x}$, $\dot{y}$, and v as a function of angle around the circle and show that all three quantities are infinite as the particle goes through the center of force.

Solution

(a) Start with position given by $r - 2a \sin\theta \ \ for \ \ 0 \le \theta \le 180°$. And have Lagrangian:

$$L = \frac{1}{2}m(\dot{r}^2 + r^2\dot{\theta}^2) - U(r) \quad with \quad \dot{r} = 2a\cos\theta\,\dot{\theta}.$$

Then,

$$\frac{d}{dt}\left(\frac{\partial L}{\partial \dot{\theta}}\right) - \frac{\partial L}{\partial \theta} = 0 \Rightarrow M_\theta = mr^2\dot{\theta} = \text{const. of motion}$$

Use $r^2 + r^2\dot{\theta}^2 = 4_a^2\cos^2\theta\,\dot{\theta}^2 + 4_a^2\sin^2\theta\,\dot{\theta}^2 = 4_a^2\dot{\theta}^2$ for the "constraint" on r to identify the respective force. Similarly, we get $E = 2ma^2\dot{\theta}^2 + U(r) = $ integral of motion, so constant:

$$E = 2ma^2\frac{M_\theta^2}{(mr^2)^2} + U(r) = \frac{2a^2 M_\theta^2}{mr^4} + U(r) = \text{const}$$

Thus,

$$\frac{dE}{dr} = -\frac{8a^2 M_\theta^2}{mr^5} + \frac{dU}{dr} = 0$$

indicates that the (attractive) force is:

$$F(r) = \frac{8a^2 M_\theta^2}{mr^5}.$$

(b) $\qquad E = \frac{2a^2 M_\theta^2}{mr^4} - \int_\infty^r -\frac{8a^2 M_\theta^2}{mr^5} = 0$

(c) $\qquad T = ? \quad M_\theta = mr^2\dot{\theta} = m(4a^2)\sin^2\theta\,\frac{d\theta}{dt}$

$$dt = m(4a^2)\frac{\sin^2\theta}{M_\theta}d\theta$$

$$T = \frac{1}{M_\theta}\int_0^\pi (4a^2)\,m\sin^2\theta\,d\theta = \frac{2\pi ma^2}{M_\theta}$$

Alternatively:

$$M_\theta = mr^2\dot{\theta} = mr \cdot r\frac{d\theta}{dt} = m2\frac{dA}{dt} \quad \rightarrow \quad dt = \frac{2mdA}{M_\theta} \quad \rightarrow \quad T = \frac{2\pi ma^2}{M_\theta}$$

(d) $\quad x = r\cos\theta = 2a\sin\theta\cos\theta = a\sin 2\theta \qquad \dot{x} = 2a(\cos^2\theta - \sin^2\theta)\dot{\theta}$

$\qquad\qquad y = r\sin\theta = 2a\sin^2\theta \qquad\qquad\qquad \dot{y} = 4a\sin\theta\cos\theta\,\dot{\theta}$

So,

$$\dot{x} = (2a)(1 - 2\sin^2\theta)\dot{\theta} = 2a\left(1 - \frac{1}{2}\left(\frac{r}{a}\right)^2\right)\frac{M_\theta}{mr^2}; \quad \dot{y}$$

$$= 2r\sqrt{1 - \left(\frac{r}{a}\right)^2\frac{M_\theta}{mr^2}}$$

and

$$v = \sqrt{4a^2\{\cos^4\theta - 2\cos^2\theta\sin^2\theta + \sin^4\theta\} + 16a^2\sin^2\theta\cos^2\theta \cdot \dot{\theta}}$$
$$= 2a\dot{\theta}\sqrt{\cos^4\theta + \sin^4\theta}.$$

Exercise 3.23. Particle in central harmonic potential.
A particle of mass m moves in central harmonic potential $V(r) = (1/2)kr^2$ with a positive spring constant k. (a) Use the effective potential to show that all orbits are bound and that E_{min} must exceed $\sqrt{kl^2/m}$. (b) Verify that the orbit is a closed ellipse with the origin at the center. If the relation $E/E_{min} = \cosh\xi$ defines the quantity ξ, show that the orbital parameters for a,b, and eccentricity. Discuss the limiting case $E \to E_{min}$ and $E \gg E_{min}$. (c) Show that the period is independent of E and l.

3.8 Small oscillations about stable equilibria

So far we've considered basic orbital mechanics and gotten the classic orbital result of an ellipse (with circle as special case). But just how stable is this idealized result for more realistic systems where there might be the occasional outside interaction nudging things? Just how stable are these solutions in 'reality'? It turns out this is a question having to do with small oscillations (to be described in detail in this section) and of overall stability (to be described in Ch. 6, where dynamics is described in phase space, and in the formalism described there the criteria for stability can be ascertained more readily). Note that broadening the class of solutions to allow for small perturbations is the first step to having a general mechanics solution but just how far can this be taken? The answer, also to follow in a later section, is up to the "boundary of chaos", which it reaches in a distinctive way, giving rise to universal constants, including C_∞ with its possibly special relation to alpha (details in [45]).

So let's consider small oscillation in the case of the circular orbit. In the potential we are in a situation where we are already at the minimum of the potential (unchanging over time). If we nudge this configuration, we see that we will experience a potential environment dominated by the potential in the neighborhood of the equilibrium, and since it is at a

63

minimum (required of equilibrium in systems in general, so this discussion generalized to those cases as well) then there is no first order term, only second to next higher order:

$$U(r) - U(r_{min}) \cong \frac{1}{2}k(r - r_{min})^2 + higher\ order\ terms$$

(Eqn. 3-39)

If we now focus on the small displacement $x = r - r_{min}$ and drop the constant $U(r_{min})$ term, we have the classic spring oscillator Lagrangian in variable x:

$$L = \frac{1}{2}m\dot{x}^2 - \frac{1}{2}kx^2$$

(Eqn. 3-40)

For which the Euler-Lagrange equations give the second order equation of motion:

$$m\ddot{x} + kx = 0 \quad \rightarrow \quad \ddot{x} + \omega^2 x = 0, \quad where\ \omega^2 = \frac{k}{m}.$$

(Eqn. 3-41)

Since convention is to speak of positive frequencies in this context, take the positive root: $\omega = \sqrt{k/m}$. The general solution for the differential equation is then: $x(t) = a\ \cos(\omega t) + b\ \sin(\omega t)$. Thus, the 1-D classic spring has two independent oscillations possible. Boundary conditions often reduce to one independent oscillation degree of freedom. Such as for the circular orbit with small oscillation problem, where orbital angular momentum is modified by the small oscillation (typically), where boundary condition selection is for spring oscillation that translates onto a wave propagation about the equilibrium circular orbit in the same orientation as the system angular momentum, giving a net system angular moment that is greater, or the opposite, with net angular momentum less. Suppose this then selects for a solution with just one of the oscillations consistent, choosing for convenience $x(t) = a\ \cos(\omega t)$, we then have:

$$E = \frac{1}{2}m\omega^2 a^2 \propto (amplitude)^2.$$

(Eqn. 3-42)

So, the system frequency is not dependent on amplitude but the system energy goes as amplitude squared. Note that the 1-D spring oscillation equation of motion can be rewritten as:

$$\frac{d^2x}{dt^2} + \omega^2 \frac{d^2x}{dX^2} = 0,$$

(Eqn. 3-43)

where the two solution classes are now captured in the form:

64

$$x(t, X) = a \, \cos(\omega t - X) + b \cos(\omega t + X).$$

(Eqn. 3-44)

Closely related to this is the 1-D (partial differential) wave equation for vibrations on string $y(t, X)$:

$$\frac{\partial^2 y}{\partial t^2} - \omega^2 \frac{\partial^2 y}{\partial X^2} = 0,$$

where the two independent solution classes are now captured in the form (D'Alembert [7]) :

$$y(t, X) = f(\omega t - X) + g(\omega t + X).$$

For both the 1-D oscillator and the 1-D string vibration, boundary conditions impact the assessment of functional degrees of freedom available.

3.8.1 Driven Systems

Now that we understand the 'natural' oscillations of the system, what if we repeatedly exert a force on the system (still remaining within the approximation of small oscillations)? By remaining within the regime of small oscillations we must have a sufficiently weak potential, and this being the case we can expand it to lowest order in displacement of the system from its equilibrium. Thus, in addition to the spring restoring force from potential energy $\frac{1}{2}kx^2$ we now have

$$U_{external}(x, t) \cong U_{ext}(0, t) + x[\partial U_{ext}/\partial x]_{x=0}$$

(Eqn. 3-45)

Dropping the term with no x dependence and writing force $F(t) = -[\partial U_{ext}/\partial x]_{x=0}$ we then get the Lagrangian for the driven oscillator:

$$L = \frac{1}{2}m\dot{x}^2 - \frac{1}{2}kx^2 + xF(t).$$

(Eqn. 3-46)

This gives rise to the differential equation:

$$\ddot{x} + \omega^2 x = \frac{F(t)}{m},$$

(Eqn. 3-47)

whose general solution can be obtained in the usual way of inhomogeneous differential equations by building off of the solutions to the homogenous differential equation. In this instance, suppose this is written as general solution $x(t) = x_{hom}(t) + x_{inhom}(t)$, where

65

$x_{hom}(t) = a \cos(\omega t + \alpha)$ like before, with $\{a, \alpha\}$ determined by boundary conditions. To calculate the $x_{inhom}(t)$ part, let's consider external forces that are periodic drivers (summation over such can then, by Fourier transform completeness, model any time-varying external force):

$$F(t) = f \cos(\gamma t + \beta).$$

(Eqn. 3-48)

If we guess a solution $x_{inhom}(t) = b \cos(\gamma t + \beta)$, we find it works for $b = f/m(\omega^2 - \gamma^2)$, thus we have for our overall solution:

$$x(t) = a \cos(\omega t + \alpha) + \left[\frac{f}{m(\omega^2 - \gamma^2)}\right] \cos(\gamma t + \beta).$$

(Eqn. 3-49)

Notice that this solution consists of a part oscillating at the system's natural frequency and a part oscillating at the driver frequency of the force. Notice also that something special happens if the driving frequency matches the system's natural frequency. This is the phenomenon of resonance.

To examine what happens at resonance we want to have a form for taking the limit $\gamma \to \omega$. For this we need the second term to be in a form amenable to using L'Hopital's rule. By simply breaking a piece of the first term and shifting its phase term as needed (all valid within the first order small oscillation approximation) we can simply rewrite:

$$x(t) = a' \cos(\omega t + \alpha) + \left[\frac{f}{m(\omega^2 - \gamma^2)}\right][\cos(\gamma t + \beta) - \cos(\omega t + \beta)],$$

(Eqn. 3-50)

and we get:

$$\lim_{\gamma \to \omega} x(t) = a' \cos(\omega t + \alpha) + \left[\frac{ft}{2m\omega}\right][\sin(\omega t + \beta)].$$

(Eqn. 3-51)

As can be seen, the familiar instability at resonance shows up in the second term, which grows linearly in time (soon violating the small oscillation assumptions). Systems often break when driven at resonance because they are able to efficiently absorb driver energy sufficient to not only violate the small oscillation assumptions (and receptivity to further driver energy absorption) but also sufficient to break a system constraint. Note: this is how a parked car can be shifted around by a small group of people pushing periodically on the car ('bouncing' without 'lifting') if the suspension is driven at resonance and lateral shoves made when at a suspension-bounce highpoint.

Let's now consider systems with more than one degree of freedom. Generally the low order terms in the potential expression in the displacements will involve cross terms. Even so, generally coordinates can be sought to decouple into a low order potential with no cross terms (known as "normal coordinates"), and the system with N degrees of freedom thereby decouples into N 1-D oscillations as already examined.

Following the notation of [27] let's consider U to be a function of multiple coordinates. We are interested in expansions of this potential with small displacements from its minimum (since assuming equilibrium with small oscillation). Using the freedom to shift the energy scale, we choose the minimum potential to be at zero, and have for potential up to quadratic terms (no linear terms since at minimum):

$$U = \frac{1}{2} \sum_{i,k} K_{ik} x_i x_k,$$

where the x's are the coordinate displacements from the minimum of the potential. Similarly, the kinetic term in generalized coordinates will still be quadratic in the velocities, but the coefficient will generally have coordinate dependence:

$$T = \frac{1}{2} \sum_{i,k} m(x_i, x_k) \dot{x}_i \dot{x}_k \cong \frac{1}{2} \sum_{i,k} m_{ik} \dot{x}_i \dot{x}_k,$$

where the latter approximation, with constant inertia matrix m_{ik} is obtained when taking the lowest order term in the generalized inertia function $\sum_{i,k} m(x_i, x_k)$ (consistent with the small displacement or small oscillation scenarios). The Lagrangian is thus:

$$L = \frac{1}{2} \sum_{i,k} (m_{ik} \dot{x}_i \dot{x}_k - K_{ik} x_i x_k),$$

and the resulting Euler-Lagrange equations:

$$\sum_k (m_{ik} \ddot{x}_k + K_{ik} x_k) = 0.$$

Consider as possible solution displacements in the generalized coordinates having different magnitudes but same the frequency: $x_k = A_k \exp i\omega t$. Substituting, we must now solve:

$$\sum_k (-\omega^2 m_{ik} + K_{ik}) A_k = 0 \quad \rightarrow \quad det|-\omega^2 m_{ik} + K_{ik}| = 0,$$

Thus, we set the determinant equal to zero, resulting in a characteristic equation of degree "N" (the number of generalized coordinates). The solutions $\{w_\alpha\}$ are the characteristic frequencies of the system. This suggests a general solution for each generalized coordinate displacement to consist of a sum over all of the characteristic frequencies (staying consistent with the notation of [27]):

$$x_k = \sum_\alpha \Delta_{k\alpha} \theta_\alpha \; ; \quad \theta_\alpha = \text{Re}[C_\alpha \exp i w_\alpha t],$$

(Eqn. 3-52)

where C_α are arbitrary complex constants, and the $\Delta_{k\alpha}$'s are the minors of the determinant associated with each of the characteristic frequencies w_α (assuming all of the w_α are different). Thus, the time variation of each coordinate of the system is a superposition of N simple periodic oscillators (with arbitrary amplitudes and phases but N definite frequencies). For simplicity, let's continue to assume all of the w_α are different and simply substitute $x_k = \sum_\alpha \Delta_{k\alpha} \theta_\alpha$, from which we get N decoupled equations upon substitution into the Lagrangian (e.g., using the characteristic frequencies we simultaneously diagonalize both kinetic and potential terms, aside from an inertial factor I_α for each frequency contribution):

$$L = \frac{1}{2} \sum_\alpha I_\alpha (\dot{\theta}_\alpha{}^2 - w_\alpha{}^2 \theta_\alpha{}^2),$$

(Eqn. 3-53)

which requires coordinate rescaling to arrive at the convention for normal coordinates that their kinetic term have a coefficient of 1/2. Thus $\theta_\alpha \rightarrow \theta_\alpha / \sqrt{I_\alpha}$, and if force is present the revised Lagrangian becomes:

$$L = \frac{1}{2} \sum_\alpha (\dot{\theta}_\alpha{}^2 - w_\alpha{}^2 \theta_\alpha{}^2) + \sum_\alpha \sum_k \frac{F_k(t)}{\sqrt{I_\alpha}} \Delta_{k\alpha} \theta_\alpha.$$

(Eqn. 3-54)

Thus, the use of normal coordinates makes possible the reduction of a forced oscillation in a system with more than one degree of freedom to a series of one-dimensional forced oscillator problems.

3.8.2 Multi-modal and locked-modal small oscillation examples
Example 3.24. Pendulum suspended from the rim of a cylindrical disk.
A simple pendulum is suspended from the rim of a cylindrical disk as
shown in Fig. 3.9. The pendulum has length l and a mass m. The disk has
a radius $r = l/2$, with a mass $M = 2m$, and can rotate freely about an
axis through its center. Find the normal modes and frequencies in the
small oscillation approximation.

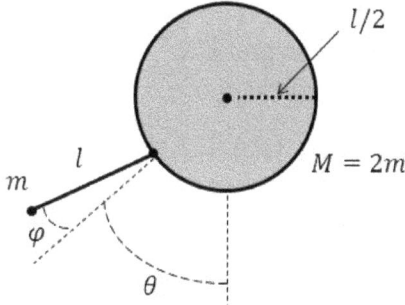

Fig. 3.9.

To obtain the Lagrangian we first need the moment of inertia for a solid
disk:

$$I = \int_0^r \rho r^2 (2\pi r) dr = 2\pi \rho \frac{r^4}{4}, \qquad where \ \rho(\pi r^2) = M,$$

thus,

$$I = \frac{1}{2} M r^2 = \frac{1}{2} (2m)(\frac{l}{2})^2 = \frac{1}{4} m l^2.$$

For the angular coordinate of the disk rotation we have θ, with angular
frequency $\omega = \dot\theta$. Let's now consider the coordinates of the pendulum
bob:

$$y = \frac{l}{2} cos\theta + l cos(\theta + \varphi) \quad and \quad x = \frac{l}{2} sin\theta + l sin(\theta + \varphi)$$

with time derivative:

$$\dot y = -\left\{ \frac{l}{2} sin\theta \dot\theta + l sin(\theta + \varphi)(\dot\theta + \dot\varphi) \right\} \quad and \quad \dot x$$

$$= \left\{ \frac{l}{2} cos\theta \dot\theta + l cos(\theta + \varphi)(\dot\theta + \dot\varphi) \right\}.$$

The kinetic terms are thus:

$$T = \frac{1}{2}I\omega^2 + \frac{1}{2}m(\dot{x}^2 + \dot{y}^2)$$

$$= \frac{1}{2}\left(\frac{1}{4}ml^2\right)\dot{\theta}^2$$

$$+ \frac{1}{2}m\left\{\left(\frac{l}{2}\dot{\theta}\right)^2 + [l(\dot{\theta} + \dot{\varphi})]^2 + l^2\dot{\theta}(\dot{\theta} + \dot{\varphi})\cos\varphi\right\}$$

The potential term is:

$$U = -mgy = -mgl\left(\frac{1}{2}\cos\theta + \cos(\theta + \varphi)\right).$$

Putting this together to get the Lagrangian and shifting to the small angle approximation (and dropping constants):

$$L = \frac{1}{8}ml^2\dot{\theta}^2 + \frac{1}{2}m\left\{\left(\frac{l}{2}\dot{\theta}\right)^2 + [l(\dot{\theta} + \dot{\varphi})]^2\right\} + mgl(\frac{1}{2}\left(-\frac{1}{2}\theta^2\right)$$

$$- \frac{1}{2}(\theta - \varphi)^2$$

$$= \frac{5}{4}ml^2\dot{\theta}^2 + \frac{3}{2}ml^2\dot{\theta}\dot{\varphi} + \frac{1}{2}ml^2\dot{\varphi}^2 - \frac{3}{4}mgl\theta^2 - mgl\theta\varphi - \frac{1}{2}mgl\varphi^2$$

Using the E-L relation, the equations of motion are then:

$$\frac{5}{2}ml^2\ddot{\theta} + \frac{3}{2}ml^2\ddot{\varphi} + \frac{3}{2}mgl\theta + mgl\varphi = 0$$

$$ml^2\ddot{\varphi} + \frac{3}{2}ml^2\ddot{\theta} + mgl\varphi + mgl\theta = 0$$

$$\left| \begin{matrix} \left(3\left(\frac{g}{l}\right) - 5\omega^2\right) & \left(2\left(\frac{g}{l}\right) - 3\omega^2\right) \\ \left(2\left(\frac{g}{l}\right) - 3\omega^2\right) & \left(2\left(\frac{g}{l}\right) - 2\omega^2\right) \end{matrix} \right| = 0$$

$$\omega^2 = \frac{4\left(\frac{g}{l}\right) \pm \sqrt{\left(4\left(\frac{g}{l}\right)\right)^2 - 4\left(2\left(\frac{g}{l}\right)^2\right)}}{2} = \left(\frac{g}{l}\right)\{2 \pm \sqrt{2}\}$$

70

and we can now write for $\omega^2 = \left(\frac{g}{l}\right)(2 + \sqrt{2})$:

$$(v - \omega^2 m)\rho^{(1)} = \begin{pmatrix} \{3 - 5(2 + \sqrt{2})\}\left(\frac{g}{l}\right) & \{2 - 3(2 + \sqrt{2})\}\left(\frac{g}{l}\right) \\ \{2 - 3(2 + \sqrt{2})\}\left(\frac{g}{l}\right) & \{2 - 2(2 + \sqrt{2})\}\left(\frac{g}{l}\right) \end{pmatrix}\begin{pmatrix} \theta \\ \varphi \end{pmatrix}$$

$$= 0$$

$$\left(-7 - 5\sqrt{2}\right)\theta + \left(-4 - 3\sqrt{2}\right)\theta = 0$$
$$\left(-4 - 3\sqrt{2}\right)\theta + \left(-2 - 2\sqrt{2}\right)\theta = 0$$

$$\theta = -\frac{\left(4 + 3\sqrt{2}\right)\varphi}{\left(7 + 5\sqrt{2}\right)} \simeq -\frac{4.1}{7}\varphi$$

Thus:

$$\rho^{(1)} \simeq c\begin{pmatrix} 1 \\ -7/4 \end{pmatrix} \quad for \quad \omega^2 = \left(\frac{g}{l}\right)(2 + \sqrt{2})$$

Similarly, for $\omega^2 = \left(\frac{g}{l}\right)(2 - \sqrt{2})$

$$(v - \omega^2 m)\rho^{(2)} = \begin{pmatrix} \{3 - 5(2 - \sqrt{2})\}\left(\frac{g}{l}\right) & \{2 - 3(2 - \sqrt{2})\}\left(\frac{g}{l}\right) \\ \{2 - 3(2 - \sqrt{2})\}\left(\frac{g}{l}\right) & \{2 - 2(2 - \sqrt{2})\}\left(\frac{g}{l}\right) \end{pmatrix}\begin{pmatrix} \theta \\ \varphi \end{pmatrix}$$

$$= 0$$

$$\theta = \frac{\left(-4 - 3\sqrt{2}\right)\varphi}{\left(-7 - 5\sqrt{2}\right)} \simeq 4\varphi$$

$$\rho^{(2)} \simeq c\begin{pmatrix} 1 \\ 1/4 \end{pmatrix} for \quad \omega^2 = \left(\frac{g}{l}\right)(2 - \sqrt{2})$$

Let's now normalize the vectors:

$$M = m\begin{pmatrix} \frac{5}{2} & \frac{3}{2} \\ \frac{3}{2} & 1 \end{pmatrix}$$

71

$$mc^2 \left(1 \quad \frac{-7}{4}\right) \begin{pmatrix} \frac{5}{2} & \frac{3}{2} \\ \frac{3}{2} & 1 \end{pmatrix} \begin{pmatrix} 1 \\ -\frac{7}{4} \end{pmatrix} = mc^2 \left(1 \quad \frac{-7}{4}\right) \begin{pmatrix} -\frac{1}{8} \\ \frac{1}{4} \end{pmatrix}$$

$$= mc^2 \left(-\frac{1}{8} + \frac{7}{16}\right) = mc^2 \left(\frac{5}{16}\right)$$

$$c \simeq \frac{4}{\sqrt{5m}}$$

$$\vec{\rho}^{(1)} = \frac{4}{\sqrt{5m}} \begin{pmatrix} 1 \\ -7/4 \end{pmatrix}$$

Similarly, we obtain for the other mode:

$$c \simeq \frac{4}{\sqrt{53m}}$$

$$\vec{\rho}^{(2)} = \frac{4}{\sqrt{53m}} \begin{pmatrix} 1 \\ 1/4 \end{pmatrix}$$

Thus, the normal modes combine to give position by:

$$\vec{x} = \frac{4}{\sqrt{5m}} \begin{pmatrix} 1 \\ -7/4 \end{pmatrix} \left\{ c_1 \cos\left(\sqrt{(2 + \sqrt{2})\left(\frac{g}{l}\right)}\, t\right) \right.$$
$$\left. + d_1 \sin\left(\sqrt{(2 + \sqrt{2})\left(\frac{g}{l}\right)}\, t\right) \right\}$$

$$+ \frac{4}{\sqrt{53m}} \begin{pmatrix} 1 \\ 1/4 \end{pmatrix} \left\{ c_2 \cos\left(\sqrt{(2 - \sqrt{2})\left(\frac{g}{l}\right)}\, t\right) \right.$$
$$\left. + d_2 \sin\left(\sqrt{(2 - \sqrt{2})\left(\frac{g}{l}\right)}\, t\right) \right\}$$

Exercise 3.24. Instead of a solid disk, have a hoop (same mass). Repeat the analysis.

Example 3.25. Two small beads on a circular wire.
For the next example, consider two small beads of mass m and charge e that move without friction on a circular wire of radius a. At t=0, the beads are diametrically opposite to one another. If bead 2 is initially at rest and bead 1 initially has speed:

$$v \ll \sqrt{\left(\frac{e^2}{ma}\right)},$$

for small oscillations, find the position of bead 1 at time t.

First, let's write the Lagrangian where the coordinates are simply the angular position of the beads:

$$L = \frac{1}{2}m\left(a^2\dot{\theta}_1{}^2 + a^2\dot{\theta}_2{}^2\right) - U(r).$$

The potential is due to the Coulomb force, so

$$F = \frac{-e^2}{r^2} \implies U = \frac{e^2}{r}.$$

Now to compute the distance r between the charges. Start by defining the angular separation between the beads: $\alpha = \theta_2 - \theta_1$ and considering the alignment of axis such that bead one is at the bottom of the wire and at the origin and bead two has

$$x = a\sin\alpha \quad and \quad y = a(1 - \cos\alpha) \quad and \quad r = a\sqrt{2(1 - \cos\alpha)}$$
$$= 2a\sin\frac{\alpha}{2}.$$

We can now write the Lagrangian as:

$$L = \frac{1}{2}ma^2\left(\dot{\theta}_1{}^2 + \dot{\theta}_2{}^2\right) - \frac{e^2}{2a\sin\frac{\alpha}{2}}$$

$$= \frac{1}{2}ma^2\left(\dot{\alpha}^2 + 2\dot{\theta}_1\dot{\alpha} + 2\dot{\theta}_1{}^2\right) - \frac{e^2}{2a\sin\frac{\alpha}{2}}$$

For small oscillations we want $\alpha = \pi + \eta$, where η is small (zero at the minimum potential), and since we have $\sin\left(\frac{\pi}{2} + \frac{\eta}{2}\right) = \cos\left(\frac{\eta}{2}\right)$ we then get:

$$L = \frac{1}{2}ma^2\left(\dot{\eta}^2 + 2\dot{\theta}_1\dot{\eta} + 2\dot{\theta}_1{}^2\right) - \frac{e^2}{2a\sin\frac{2}{\eta}}$$

The equations of motion then follow from the E-L relation, $\frac{d}{dt}\left(\frac{\partial L}{\partial \dot{q}}\right) - \frac{\partial L}{\partial q} = 0$, to give:

$$\frac{1}{2}ma^2(2\ddot{\eta} + 4\ddot{\theta}_1) = 0 \implies \ddot{\theta}_1 = -\frac{1}{2}\ddot{\eta}$$

$$\frac{1}{2}ma^2(2\ddot{\eta} + 2\ddot{\theta}_1) + \frac{e^2}{2a}\left(\frac{-\left(-\sin\left(\frac{\eta}{2}\right)\frac{1}{2}\right)}{\cos^2\left(\frac{\eta}{2}\right)}\right) = 0$$

And approximating for small η:

$$\ddot{\eta} + \frac{e^2}{2ma^3}\left(\frac{\eta}{2}\right) = 0,$$

and the frequency of small oscillations for the system is:

$$\omega^2 = \frac{e^2}{4ma^3}.$$

At time t=0 we have $\alpha = \pi \Rightarrow \eta = 0$. Writing the general solution for the given frequency of oscillation:

$$\eta = Bsin(\omega t).$$

Now, at $t = 0$ we have $v_2 = v$, $v_1 = 0$, so:

$$v_2 = a\dot{\theta}_2 = v, \quad and \quad \dot{\eta} = \dot{\alpha} = \dot{\theta}_2 - \dot{\theta}_1 = \dot{\theta}_2 = \frac{v}{a} \quad at\ t = 0$$

$$\dot{\eta} = B\omega cos(\omega t)\Big|_{t\,=\,0} = \left(\frac{v}{a}\right) \quad \rightarrow \quad B = \frac{v}{a\omega}$$

Thus, $\eta = \frac{v}{a\omega}sin(\omega t)$, and we can write

$$\ddot{\theta}_1 = -\frac{1}{2}\ddot{\eta} \quad \rightarrow \quad \frac{d}{dt}\left(\dot{\theta}_1 + \frac{1}{2}\dot{\eta}\right) = 0 \quad \rightarrow \quad \dot{\theta}_1 + \frac{1}{2}\dot{\eta} = \frac{v}{2a}$$

and

$$\dot{\theta}_1 = \frac{v}{2a} - \frac{1}{2}\dot{\eta} \quad \rightarrow \quad \theta_1 = \frac{v}{2a}t - \frac{v}{2a\omega}sin(\omega t) + \theta_0$$

where θ_0 is the initial angle for θ_1. Thus,

$$\theta_1 = \frac{v}{2a}\left\{t - \frac{sin(\omega t)}{\omega}\right\} + \theta_0, \quad \omega = \sqrt{\frac{e^2}{4ma^3}}$$

Exercise 3.25. Have the two beads be at rest, positioned 175 degrees apart, and release. For small oscillations, find positions of the beads at time t.

Example 3.26. Pendulum within rolling hoop.

Now consider a thin cylindrical hoop of radius R and mass M which rolls without slipping on a rough horizontal surface (Fig 3.10). A physical pendulum of mass m is mounted onto the axis of the cylinder by means of an arrangement of spokes of negligible mass converging at the origin and providing a pendulum mount that is free to rotate freely about the cylindrical axis. The center of mass of the pendulum is at a distance h from the cylindrical axis, and its radius of gyration is k. For small

74

oscillations about the equilibrium position obtain the period of oscillation in terms of the aforementioned variables.

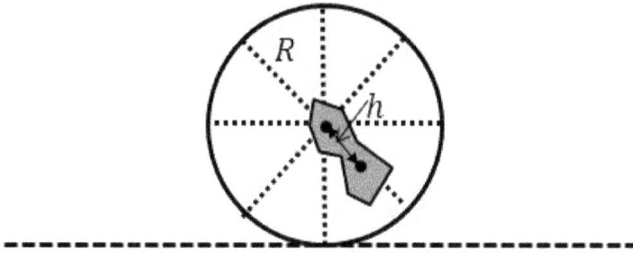

Fig. 3.10.

The kinetic energy of the hoop is:
$$T_h = \frac{1}{2}I_h\omega_h^2 + \frac{1}{2}Mv_h^2, \quad where \quad I_h = MR^2 \quad and \quad \omega_h = \dot{\theta}, \quad v_h = R\dot{\theta}$$

The kinetic energy of the pendulum is:
$$T_p = \frac{1}{2}I_{p(cm)}\omega_p^2 + \frac{1}{2}mv_p^2$$

The pendulum's moment of inertia is given by the parallel axis theorem:
$$I = I_{cm} + mh^2 \quad \rightarrow \quad I_{p(cm)} = mk^2 - mh^2$$

Writing the position of the pendulum in Cartesian coordinates:
$$x = hsin\varphi \quad and \quad y = -hcos\varphi,$$
with time derivatives:
$$\dot{x} = hcos\varphi\dot{\varphi} \quad and \quad \dot{y} = hsin\varphi\dot{\varphi}.$$
For the velocities we can then write:
$$\omega_p = \dot{\varphi} \quad and \quad v_T = |\vec{v}_h + \vec{v}_p| = \sqrt{(v_h + h\dot{\varphi}cos\varphi)^2 + (h\dot{\varphi}sin\varphi)^2}$$

The total velocity of the pendulum center of mass is thus
$$v_T^2 = v_h^2 + (h\dot{\varphi})^2 + 2v_h(h\dot{\varphi})cos\varphi$$

and the pendulum potential energy is:
$$U = -mghcos\varphi.$$
We can now write the Lagrangian:

75

$$L = \frac{1}{2}MR^2\dot{\theta}^2 + \frac{1}{2}M(R\dot{\theta})^2 + \frac{1}{2}(mk^2 - mh^2)\dot{\varphi}^2$$
$$+ \frac{1}{2}m\{v_h{}^2 - (h\dot{\varphi})^2 + 2v_h(h\dot{\varphi})\cos\varphi\} + mghcos\varphi$$

and now switching to the small oscillation formalism (dropping 3rd order terms and higher):

$$L = MR^2\dot{\theta}^2 + \frac{1}{2}(mk^2 - mh^2)\dot{\varphi}^2 + \frac{1}{2}m\{(R\dot{\theta})^2 + (h\dot{\varphi})^2 + 2(R\dot{\theta})(h\dot{\varphi})\}$$
$$- \frac{1}{2}mgh\varphi^2$$
$$= \left(MR^2 + \frac{1}{2}mR^2\right)\dot{\theta}^2 + \frac{1}{2}mk^2\dot{\varphi}^2 + mRh\dot{\theta}\dot{\varphi} - \frac{1}{2}mgh\varphi^2$$

We can now get the equations of motion using the E-L equations:

$$\theta \text{ equation:} \quad 2\left(MR^2 + \frac{1}{2}mR^2\right)\ddot{\theta} + mRh\ddot{\varphi} = 0$$
$$\Longrightarrow \quad \frac{d}{dt}\{(2M + m)R^2\dot{\theta} + mhR\dot{\varphi}\} = 0$$

Thus we get $\ddot{\theta} = -\frac{mRh\ddot{\varphi}}{(2M+m)R^2}$, which we use in the other equation:

$$\varphi \text{ equation:} \quad mk^2\ddot{\varphi} + mhR\ddot{\theta} + mgh\varphi = 0$$

rewriting after substitution:

$$\left\{mk^2 - \frac{m^2h^2}{(2M + m)}\right\}\ddot{\varphi} + mgh\varphi = 0$$

$$\omega^2 = \frac{mgh}{mk^2 - \dfrac{m^2h^2}{(2M + m)}} \quad \rightarrow \quad \omega = \sqrt{\frac{g}{h}\left\{\left(\frac{k}{h}\right)^2 - \frac{m}{(2M + m)}\right\}^{-1}}$$

And as $M \to \infty$ the hoop becomes ignorable and the frequency becomes $\omega = \sqrt{\frac{gh}{k^2}}$ as expected. For the period we then get:

$$T = \frac{2\pi}{\omega} = 2\pi\sqrt{\frac{k^2}{gh}}\sqrt{1 - \left(\frac{h}{k}\right)^2\frac{m}{(2M + m)}}.$$

Note how there is no R-dependence in the solution.

Exercise 3.26. Replace the hoop with a solid disk. (Ignore effects of thickness.)

Example 3.27. A particle in a potential $V(\vec{r}) = V_0 \log r$.
A particle of mass m moves in a potential $V(\vec{r}) = V_0 \log r$. Let Ω be the frequency of a circular orbit at r=R, and let ω be the frequency of small radial oscillations about that circular orbit. Find ω/Ω.

Starting with the Lagrangian in polar coordinates:
$$L = \frac{1}{2}m(\dot{r}^2 + r^2\dot{\theta}^2) - V(\vec{r}) = \frac{1}{2}m(\dot{r}^2 + r^2\dot{\theta}^2) - V_0 \log r$$

From the E-L equations for the θ coordinate we get:
$$\frac{d}{dt}(mr^2\dot{\theta}) = 0 \rightarrow mr^2\dot{\theta} = l.$$
For the r coordinate we get:
$$m\ddot{r} - mr\dot{\theta}^2 + \frac{V_0}{r} = 0 \rightarrow \ddot{r} - \frac{l^2}{m^2r^3} + \frac{V_0}{m}\frac{1}{r} = 0$$

For circular orbits $r = R$ we get $R^2 = \frac{l^2}{mV_0}$, or:
$$R = \frac{l}{\sqrt{mV_0}}.$$
The period of the circular orbit is given by integrating $mr^2\dot{\theta} = l$ to get $mr^2(\frac{2\pi}{T}) = l$ over one cycle. Thus, the period is $T = mr^2(\frac{2\pi}{l})$. Relating the period to the frequency, then have:
$$\Omega = \frac{l}{mR^2} = \frac{V_0}{l}$$

Now let's consider small radial oscillations:
$$r = R + \eta \rightarrow \ddot{\eta} - \frac{l^2}{m^2(R+\eta)^3} + \frac{V_0}{m}\frac{1}{(R+\eta)} = 0$$

which simplifies for small η to be:
$$\ddot{\eta} + \eta\left(\frac{V_0^2}{l^2}\right)2 = 0 \implies \omega = \frac{V_0}{l}\sqrt{2}.$$
Thus, the ratio of frequencies is:
$$\frac{\omega}{\Omega} = \sqrt{2}.$$

Exercise 3.27. Try as in Ex. 3.27, but with $V(\vec{r}) = -V_0/r$

Example 3.28. Massless hoop with pendulum.

A massless hoop of radius 2l rolls without slipping on a flat floor (Fig. 3.11). Attached to the loop is a rod of length 2l and mass m that can swing freely in the plane of the hoop. Find the frequency of the oscillatory mode for small oscillations about the equilibrium position shown.

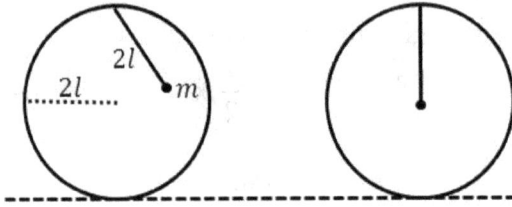

Fig. 3.11.

Let's use the angle θ to specify the displacement from the equilibrium position for the support point, then $\omega_1 = \dot\theta$ and the non-slip condition relates this to the horizontal velocity of the hoop: $v_h = 2l\omega_1\dot\theta$.

The moment of inertia for the rod is:

$$I = \frac{1}{3}mR^2 = \frac{1}{3}(m)(2l)^2 = \frac{4}{3}ml^2$$

Let's now express the position of the rod support point in Cartesian coordinates:

$$x_s = (2l)\sin\theta \quad and \quad y_s = 2l + (2l)\cos\theta,$$

for which the coordinate time-derivatives are:

$$\dot x_s = 2l\cos\theta\dot\theta \quad and \quad \dot y_s = -2l\sin\theta\dot\theta.$$

Let's now express the position of the rod center of mass, respective to support point, by the angle φ:

$$x = (l)\sin\varphi \quad and \quad y = -(l)\cos\varphi,$$

for which the coordinate time-derivatives are:

$$\dot x = l\cos\theta\dot\varphi \quad and \quad \dot y = -l\sin\varphi\dot\varphi.$$

We can now write the kinetic energy:

$$v = |\vec{v_s} + \vec{v_{cm}}| = \sqrt{((v_s)_x + \dot x)^2 + ((v_s)_y + \dot y)^2}$$

after substitutions:

$$v^2 = (v_h + (2l)\omega_1\cos\theta)^2 + 2(v_h + (2l)\omega_1\cos\theta)\dot x + \dot x^2$$
$$+ (-(2l)\omega_1\sin\theta)^2 - 2((2l)\omega_1\sin\theta)\dot y + \dot y^2$$

78

$$v^2 = 2[(2l)\omega_1]^2 + 2[(2l)\omega_1]cos\theta + 2(2l)\omega_1(1 + cos\theta)\dot{x}$$
$$- 2(2l)\omega_1 sin\theta\dot{y} + (l\dot{\varphi})^2$$

Thus,

$$T = \frac{1}{2}I\omega^2 + \frac{1}{2}mV^2$$

$$T = \frac{1}{2}\left(\frac{4}{3}ml^2\right)\dot{\varphi}^2$$

$$+ \frac{1}{2}m\left\{2\left(2l\dot{\theta}\right)^2(1 + cos\theta) + 2\left(2l\dot{\theta}\right)(1 + cos\theta)\dot{x}\right.$$
$$\left. - 2\left(2l\dot{\theta}\right)sin\theta\dot{y} + (l\dot{\varphi})^2\right\}$$

The potential energy is given by:

$$U = -mgy_{cm} = -mg(y_s + y) = -mg\{2l + 2lcos\theta - lcos\varphi\}$$

Putting this together to get the Lagrangian and assuming small angles:

$$L = T - U = \frac{2}{3}ml^2\dot{\varphi}^2 + 2m\left(2l\dot{\theta}\right)^2 + 2m\left(2l\dot{\theta}\right)(l\dot{\varphi}) + (l\dot{\varphi})^2 - mgl\theta^2$$

$$+ mgl\left(\frac{\varphi^2}{2}\right)$$

We can now compute the equations of motion:

$$\theta: \quad 4m(2l)^2\ddot{\theta} + m(2l)^2\ddot{\varphi} + 2mgl\theta = 0$$

$$\varphi: \quad \frac{1}{3}m(2l)^2\ddot{\varphi} + m(2l)^2\ddot{\theta} - mgl\varphi = 0$$

After simplification:

$$\theta: \quad 4\ddot{\theta} + \ddot{\varphi} + \frac{g}{2l}\theta = 0$$

$$\emptyset: \quad \frac{1}{3}\ddot{\varphi} + \ddot{\theta} - \frac{g}{4l}\varphi = 0$$

Solving to get the normal mode frequencies:

$$\begin{vmatrix} \frac{g}{2l} & -\omega^2 \\ -\omega^2 & \frac{g}{4l} - \frac{1}{3}\omega^2 \end{vmatrix} = 0 \quad \rightarrow \quad \omega^2 = \left(\frac{g}{2l}\right)\left\{\frac{-5 \pm \sqrt{25 + 6}}{2}\right\}$$

and for oscillatory mode we take the $\omega^2 > 0$ root:

$$\omega^2_{osc} = \left(\frac{g}{2l}\right)\left(\frac{\sqrt{31} - 5}{2}\right).$$

Exercise 3.28. Try as in Ex. 3.28, but with hoop having mass M.

Example 3.29. Balls and springs problem.

Consider three balls B, C, D, that are connected in a line BCD by two springs. Consider all motion to be along the x-axis. Consider a ball A coming from the left on a collision course with ball B. Take all four ball masses to be m. Take the two spring constants to be k. The initial grouping of three balls is at rest, while the approaching ball A is at velocity v. Let the collision occur at time=0 and assume the collision time is short comparted to $\sqrt{(m/k)}$. Find the position of ball D as a function of time.

The Lagrangian for the BCD system is simply:

$$L = \frac{1}{2}m(\dot{x}_B{}^2 + \dot{x}_C{}^2 + \dot{x}_D{}^2)$$

$$-\frac{1}{2}k([x_C - x_B]^2 + [x_D - x_C]^2)$$

$$\tilde{v} = k\begin{vmatrix} 1 & -1 & 0 \\ -1 & 2 & -1 \\ 0 & -1 & 1 \end{vmatrix} \text{ and } \tilde{m} = m\begin{vmatrix} 1 & 0 & 0 \\ 0 & 1 & 0 \\ 0 & 0 & 1 \end{vmatrix} \text{ and } |\tilde{v} - \omega^2\tilde{m}| = 0$$

Then give the determinant:

$$\begin{vmatrix} k - \omega^2 m & -k & 0 \\ -k & 2k - \omega^2 m & -k \\ 0 & -k & k - \omega^2 m \end{vmatrix} = 0$$

thus

$$m\omega^2(k - \omega^2 m)(3k - \omega^2 m) = 0$$

And, the frequencies are: $\omega = 0$; $\omega = \sqrt{k/m}$; and $\omega = \sqrt{3k/m}$, where $\omega = 0$ corresponds to translation. For mode $\omega_1 = 0$:

$$(\tilde{v} - \omega^2\tilde{m})\rho^{(1)} = \begin{pmatrix} 1 & -1 & 0 \\ -1 & 2 & -1 \\ 0 & -1 & 1 \end{pmatrix}\begin{pmatrix} x_B \\ x_C \\ x_D \end{pmatrix} = 0 \qquad \rightarrow \qquad \rho^{(1)} = c\begin{pmatrix} 1 \\ 1 \\ 1 \end{pmatrix}$$

Now to get the normalization:

$$\rho^{(1)}m\rho^{(1)} = mc^2(1 \quad 1 \quad 1)\begin{pmatrix} 1 & & \\ & 1 & \\ & & 1 \end{pmatrix}\begin{pmatrix} 1 \\ 1 \\ 1 \end{pmatrix} = c^2(3)m = 1$$

Thus

$$\rho^{(1)} = \frac{1}{\sqrt{3m}}\begin{pmatrix} 1 \\ 1 \\ 1 \end{pmatrix}$$

For mode $\omega_2 = \sqrt{\frac{k}{m}}$:

$$\begin{pmatrix} 0 & -k & 0 \\ -k & k & -k \\ 0 & -k & 0 \end{pmatrix} \begin{pmatrix} x_B \\ x_C \\ x_D \end{pmatrix} = 0 \quad \rightarrow \quad \vec{p}^{(2)} = c \begin{pmatrix} 1 \\ 0 \\ -1 \end{pmatrix} \quad \rightarrow \quad \vec{p}^{(2)}$$

$$= \frac{1}{\sqrt{2m}} \begin{pmatrix} 1 \\ 0 \\ -1 \end{pmatrix}$$

And for mode $\omega_3 = \sqrt{\dfrac{3k}{m}}$:

$$\begin{pmatrix} -2k & -k & 0 \\ -k & k & -k \\ 0 & -k & -2k \end{pmatrix} \begin{pmatrix} x_B \\ x_C \\ x_D \end{pmatrix} = 0 \quad \rightarrow \quad \vec{p}^{(3)} = c \begin{pmatrix} 1 \\ -2 \\ 1 \end{pmatrix} \quad \rightarrow \quad \vec{p}^{(2)}$$

$$= \frac{1}{\sqrt{6m}} \begin{pmatrix} 1 \\ -2 \\ 1 \end{pmatrix}$$

The general form of the solution with these three modes is:
$$\vec{x}(t) = \vec{p}^{(1)}(c_1 + d_1 t) + \vec{p}^{(2)}(c_2 \cos \omega_2 t + d_2 \sin \omega_2 t)$$
$$+ \vec{p}^{(3)}(c_3 \cos \omega_3 t + d_3 \sin \omega_3 t)$$

$$\vec{x}(0) = \begin{pmatrix} 0 \\ 0 \\ 0 \end{pmatrix} \implies c_1 = 0, c_2 = 0, c_3 = 0$$

For the velocities we start with
$$\dot{\vec{x}}(0) = \begin{pmatrix} v \\ 0 \\ 0 \end{pmatrix} = \vec{v}$$

Then,

$$\dot{\vec{x}}(0)\tilde{m}\vec{p}^{(1)} = d_1 = (v\ 0\ 0)\frac{m}{\sqrt{3m}}\begin{pmatrix} 1 \\ 1 \\ 1 \end{pmatrix} = \frac{mv}{\sqrt{3m}} \rightarrow d_1 = \frac{mv}{\sqrt{3m}}$$

$$\dot{\vec{x}}(0)\tilde{m}\vec{p}^{(2)} = \omega_2 d_2 = (v\ 0\ 0)\frac{m}{\sqrt{2m}}\begin{pmatrix} 1 \\ 0 \\ -1 \end{pmatrix} = \frac{mv}{\sqrt{2m}} \rightarrow d_2 = \frac{mv}{\sqrt{2k}}$$

$$\dot{\vec{x}}(0)\tilde{m}\vec{p}^{(3)} = \omega_3 d_3 = (v\ 0\ 0)\frac{m}{\sqrt{6m}}\begin{pmatrix} 1 \\ -2 \\ 1 \end{pmatrix} = \frac{mv}{\sqrt{6m}} \rightarrow d_3 = \frac{mv}{3\sqrt{2k}}$$

Thus,
$$\vec{x}(t) = \frac{v}{3}\begin{pmatrix} 1 \\ 1 \\ 1 \end{pmatrix}t + \frac{v}{2\omega_2}\begin{pmatrix} 1 \\ 0 \\ -1 \end{pmatrix}\sin\omega_2 t + \frac{v}{6\omega_2}\begin{pmatrix} 1 \\ -2 \\ 1 \end{pmatrix}\sin\omega_3 t$$

For ball D specifically:
$$x_D(t) = \frac{v}{3}t - \frac{v}{2\omega_2}\sin\omega_2 t + \frac{v}{6\omega_2}\sin\omega_3 t.$$

Exercise 3.29. Try as in Ex. 3.29, but with ball C having mass 2m, not m.

Example 3.30. Rods with torsional springs.
Two uniform thin rods each of mass m and length l are connected by a torsional spring and one of them has other end attached by torsional spring to a fixed point. The torsional springs have torque $= k\theta$. The free end of the outside rod is pushed by a force F. (a) What are the Euler-Lagrange equations; (b) In the small-oscillation approximation, what are the frequencies?

Solution
(a) The potential energy from the torsion springs is:
$$U = \frac{1}{2}k\left[\theta_1{}^2 + (\theta_2 - \theta_1)^2\right]$$
Note that the moment of inertia for the two rods must be treated differently as one rod has a fixed end, thus will undergo rotations about that fixed point, for which the relevant moment of inertial is
$$I_1 = \frac{1}{3}ml^2,$$
while the other rod is not fixed, so we will consider its motion in its center-of-mass (CM) frame, where the relevant moment of inertia is about the center:
$$I_2 = \frac{1}{12}ml^2.$$
We can now write the Lagrangian:
$$L = \frac{1}{2}I_1\omega_1{}^2 + \frac{1}{2}I_2\omega_2{}^2 + \frac{1}{2}M_2v_2{}^2 - U.$$
Now to get the CM velocity of the rod with free ends:
$$x = l\left(\sin\theta_1 + \frac{1}{2}\sin\theta_2\right) \quad and \quad y = l\left(\cos\theta_1 + \frac{1}{2}\cos\theta_2\right),$$
and the velocities are:
$$\dot{x} = l\left(\cos\theta_1\dot{\theta}_1 + \frac{1}{2}\cos\theta_2\dot{\theta}_2\right) \quad and \quad \dot{y} = -l\left(\sin\theta_1\dot{\theta}_1 + \frac{1}{2}\sin\theta_2\dot{\theta}_2\right)$$
Thus, the velocities are:
$$v_2{}^2 = (l\dot{\theta}_1)^2 + \left(\frac{l}{2}\dot{\theta}_2\right)^2 + l^2\dot{\theta}_1\dot{\theta}_2\{\cos\theta_1\cos\theta_2 + \sin\theta_1\sin\theta_2\}$$
and according to choice of angles:
$$\omega_1 = \dot{\theta}_1 \quad and \quad \omega_2 = -\dot{\theta}_2$$
The Lagrangian is thus:

$$L = \frac{1}{2}\left(\frac{1}{3}ml^2\right)\dot{\theta_1}^2 + \frac{1}{2}\left(\frac{1}{12}ml^2\right)\dot{\theta_2}^2$$
$$+ \frac{1}{2}m\left\{(l\dot{\theta_1})^2 + (\frac{l}{2}\dot{\theta_2})^2 + l^2\dot{\theta_1}\dot{\theta_2}\cos(\theta_2 - \theta_1))\right\} - U$$

For which the equations of motion are:

$$\theta_1: \left(ml^2 + \frac{ml^2}{3}\right)\ddot{\theta_1} + \frac{d}{dt}\left\{\frac{1}{2}ml^2\dot{\theta_2}\cos(\theta_2 - \theta_1)\right\}$$
$$- \frac{1}{2}ml^2\dot{\theta_1}\dot{\theta_2}\sin(\theta_2 - \theta_1)) + \{k\theta_1 + k(\theta_2 - \theta_1)(-1)\}$$
$$= F_1$$
$$\frac{4ml^2}{3}\ddot{\theta_1} + \frac{ml^2}{2}\left\{\ddot{\theta_2}\cos(\theta_2 - \theta_1)\right.$$
$$\left. - (\dot{\theta_2})^2\sin(\theta_2 - \theta_1)\right\} + k\{2\theta_1 - \theta_2\} = F_1$$

and

$$\theta_2: \frac{ml^2}{3}\ddot{\theta_2} + \frac{ml^2}{2}\left\{\ddot{\theta_1}\cos(\theta_2 - \theta_1) + (\dot{\theta_1})^2\sin(\theta_2 - \theta_1)\right\} + k(\theta_2 - \theta_1)$$
$$= F_2$$

where

$$F_{\theta_2} = F_y\frac{\partial y}{\partial \theta_1} = (-F)(-l\sin\theta_2) = Fl\sin\theta_2 \quad and \quad F_{\theta_1} = (-F)\frac{\partial y}{\partial \theta_1}$$
$$= Fl\sin\theta_1$$

Thus,

$$\theta_1: \frac{4}{3}ml^2\ddot{\theta_1} + \frac{ml^2}{2}\left\{\ddot{\theta_2}\cos(\theta_2 - \theta_1) - \dot{\theta_2}^2\sin(\theta_2 - \theta_1)\right\} + k\{2\theta_1 - \theta_2\}$$
$$= Fl\sin\theta_1$$

and

$$\theta_2: \frac{1}{3}ml^2\ddot{\theta_2} + \frac{ml^2}{2}\left\{\ddot{\theta_1}\cos(\theta_2 - \theta_1) - \dot{\theta_1}^2\sin(\theta_2 - \theta_1)\right\} + k\{\theta_2 - \theta_1\}$$
$$= Fl\sin\theta_2$$

(b) Now switching to small oscillations:

$$\frac{4}{3}ml^2\ddot{\theta_1} + \frac{ml^2}{2}\{\ddot{\theta_2}\} + k\{2\theta_2 - \theta_1\} - Fl\theta_1 = 0$$

and

$$\frac{1}{3}ml^2\ddot{\theta_2} + \frac{ml^2}{2}\{\ddot{\theta_1}\} + k\{\theta_2 - \theta_1\} - Fl\theta_2 = 0$$

Now to get the normal mode frequencies from evaluating the determinant:

$$\begin{vmatrix} -[2k+Fl]-\dfrac{4}{3}ml^2\omega^2 & -k-\dfrac{1}{2}ml^2\omega^2 \\[2mm] -k-\dfrac{1}{2}ml^2\omega^2 & -[-k+Fl]-\dfrac{1}{3}ml^2\omega^2 \end{vmatrix}=0$$

$$\left([-2k+Fl]+\frac{4}{3}ml^2\omega^2\right)\left([-k+Fl]+\frac{1}{3}ml^2\omega^2\right)-\left(-k-\frac{1}{2}ml^2\omega^2\right)$$
$$=0$$

When $Fl \gg k$:

$$\left(Fl+\frac{4}{3}ml^2\omega^2\right)\left(Fl+\frac{1}{3}ml^2\omega^2\right)\cong 0 \;\rightarrow\; \omega_1{}^2=-\frac{3F}{4ml}\;\;and\;\;\omega_2{}^2$$
$$=-\frac{3F}{ml}$$

When $Fl \ll k$:

$$\left(-2k+\frac{4}{3}ml^2\omega^2\right)\left(-k+\frac{1}{3}ml^2\omega^2\right)-(k+\frac{1}{2}ml^2\omega^2)^2=0$$

where the frequencies are:

$$\omega^2=\frac{3kml^2\pm\sqrt{9-\frac{28}{36}(kml^2)}}{2*\frac{7}{36}(ml^2)^2}\quad (both\;positive).$$

Exercise 3.30. Try as in Ex. 3.30, but with fixed end now free.

3.8.3 Damping
Now that we've covered free and forced oscillations, the next key phenomenological effect is damping (friction), and this finally gives us a first-order time-derivative term in the equations of motion, e.g., we now have an opposing frictional force linear in velocity ($F=-\alpha\dot{x}$):

$$m\ddot{x}+kx=-\alpha\dot{x}\quad\rightarrow\quad \ddot{x}+2\lambda\dot{x}+\omega^2 x=0, where\;\omega^2=\frac{k}{m}\;and\;2\lambda$$
$$=\frac{\alpha}{m}.$$

To solve, try the form $x=\exp(rt)$ which has roots of characteristic equation: $r_{1,2}=-\lambda\pm\sqrt{\lambda^2-\omega^2}$. Thus, $x(t)=c_1\exp(r_1t)+c_2\exp(r_2t)$ in the general solution and we have the following cases:

Case $<\omega$: _exponentially damped oscillations_
$$x(t)=a\exp(-\lambda t)\cos(\omega't+\alpha),\qquad \omega'=\sqrt{\omega^2-\lambda^2}.$$
Notice that there is a decrease in frequency since friction retards motion.

Case $= \omega$: *exponentially damped with no oscillation*
$$x(t) = (c_1 + c_2 t) \exp(-\lambda t).$$
Case $> \omega$: *Aperiodic damping*
$$x(t) = c_1 \exp(r_1 t) + c_2 \exp(r_2 t), \text{with } r_{1,2} \text{ roots real and negative.}$$

3.8.4 First encounter with the Dissipative function
Consider friction in the multidimensional case with $N>1$ degrees of freedom $F_i = -\sum_k \alpha_{ik} \dot{x}_k$. To avoid rotational instability or other statistical mechanics pathologies, we require α_{ik} to be symmetric, thus we can introduce a dissipation function $\mathcal{F}$:

$$\mathcal{F} = \frac{1}{2} \sum_{i,k} \alpha_{ik} \dot{x}_i \dot{x}_k, \text{where } F_i = -\frac{\partial \mathcal{F}}{\partial x_i}$$

(Eqn. 3-55)

Let's consider the rate of dissipation of energy in the system:

$$\frac{dE}{dt} = \frac{d}{dt}\left(\sum_i \dot{x}_i \frac{\partial L}{\partial \dot{x}_i} - L \right) = -\sum_i \dot{x}_i \frac{\partial \mathcal{F}}{\partial \dot{x}_i} = -2\mathcal{F}.$$

(Eqn. 3-56)

Thus $\mathcal{F}$ is proportional to the rate of dissipation of energy as its name suggests.

2.8.5 Forced oscillations under friction
In this section we combine both the frictional force and driving force in combination. The general form of the differential equation describing forced oscillation with damping (complex form) is:

$$\ddot{x} + 2\lambda\dot{x} + \omega^2 x = \left(\frac{F}{m}\right)\exp i\gamma t.$$

(Eqn. 3-57)

Try $x(t) = B \exp(i\gamma t)$ for the particular solution, then the characteristic equation gives us:

$$B = \frac{F}{m(\omega^2 - \gamma^2 + 2i\lambda\gamma)} = b \exp(i\delta),$$

(Eqn. 3-58)

where

$$b = \frac{F}{m\sqrt{(\omega^2 - \gamma^2)^2 + (2\lambda\gamma)^2}}, \text{and } \tan \delta = \frac{(2\lambda\gamma)}{(\omega^2 - \gamma^2)}.$$

(Eqn. 3-59)

Adding the particular solution to the general solution for the homogeneous equation (and taking $\omega > \lambda$ for definiteness in what follows), and taking the real part as our solution, we have:

$$x(t) = a \exp(-\lambda t) \cos(\omega t + \alpha) + b \cos(\gamma t + \delta),$$

(Eqn. 3-60)

and after sufficient time, there is just $x(t) \cong b \cos(\gamma t + \delta)$.

Near resonance, $\gamma = \omega + \epsilon$, suppose also that $\lambda \ll \omega$, then

$$b = \frac{F}{2m\omega\sqrt{\epsilon^2 + \lambda^2}}, \, and \, \tan \delta = \frac{\lambda}{\epsilon}.$$

(Eqn. 3-61)

The phase difference δ between the oscillation and the external force is always negative. Far from resonance, $\gamma < \omega$: $\delta \to 0$; and $\gamma > \omega$: $\delta \to -\pi$. While passing thru resonance $\gamma = \omega$: $\delta \to -\frac{1}{2}\pi$. In the absence of friction, the phase of the forced oscillation changes discontinuously by π at $\gamma = \omega$; when friction added, the discontinuity smooths out.

Once steady state motion is achieved, $x(t) \cong b \cos(\gamma t + \delta)$, energy absorbed from the external force matches that dissipated in friction. We have the rate of dissipation due to friction previously as $-2\mathcal{F}$, where $\mathcal{F} = \frac{1}{2}\alpha\dot{x}^2 = \lambda m b^2 \gamma^2 \sin^2(\gamma t + \delta)$, with time average: $2\bar{\mathcal{F}} = \lambda m b^2 \gamma^2$. Thus the energy absorbed per unit time is $\lambda m b^2 \gamma^2$. Now if we want the integral of the energy absorbed at all driving frequencies, the absorption will be dominated by the frequencies near resonance, for which the integral approximates as $\pi F^2/4m$.

Note, in this analysis we are considering the spring or pendulum with only a linear restoring force. For the pendulum in the small angle approximation, however, such is the case, where the force due to gravity term is $-mg\sin(\theta) \cong -mg\theta$. When we return to the damped driven oscillator without this approximation later, we will see that chaotic motion becomes ubiquitous among the possible motions elicited.

Before moving on from the topic of dissipation, and to get a glimpse of the phase diagram representation used in the Hamiltonian approach to be discussed next, let's consider the system:

$$m\ddot{x} + \gamma\dot{x} + \frac{dU}{dx} = 0,$$

(Eqn. 3-62)

86

when the potential is a double well. In Fig. 3.12 is shown a sketch of the potential, of the system phase diagram when $\gamma = 0$ (no dissipation), and of the system phase diagram when $\gamma \neq 0$. For the system with dissipation, we see that there is a decaying spiral that selects a well to localize to when the energy dissipates to the level of the separatrix.

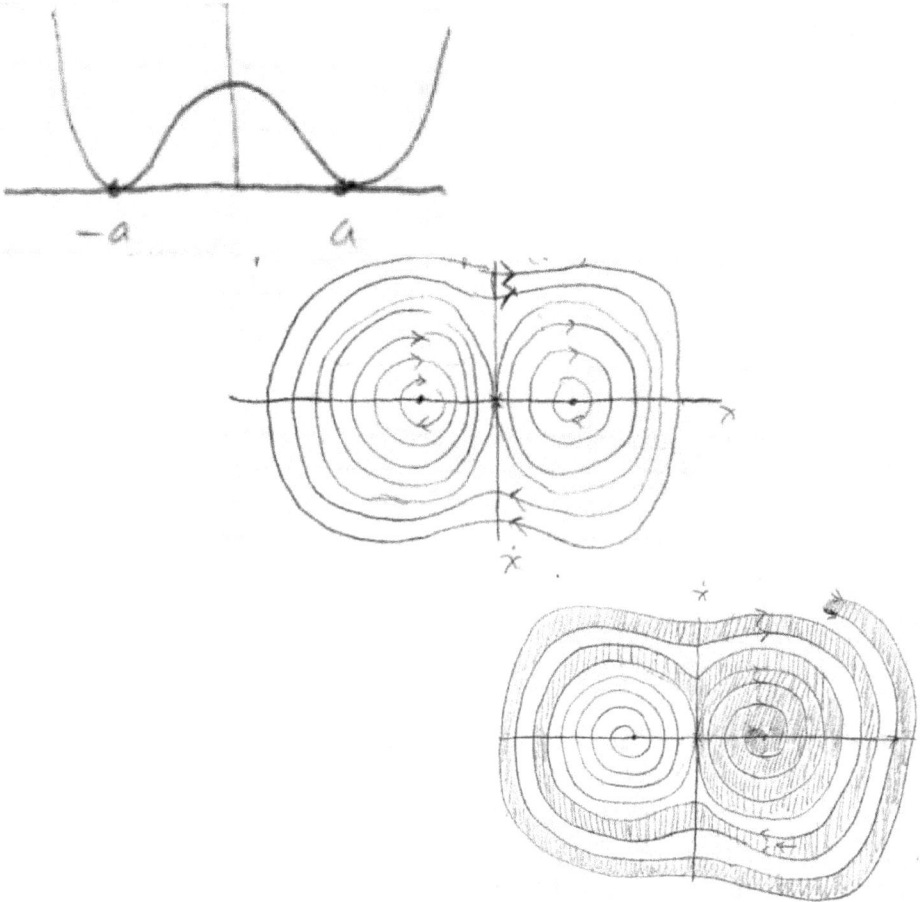

Fig. 3.12. Left: a sketch of a double-well potential; Middle: Sketch of phase diagram with no dissipation; Phase diagram with dissipation (and eventual settling into the right well).

2.8.6 Parametric resonance

Instead of an external force, let's now consider modulations of the system parameters themselves (system isn't closed). For an external force driving the system at resonance we found linear growth over time in system displacement from equilibrium. For parametric resonance we will see this growth at resonance is *exponential*, where the growth is multiplicative, but this also means this resonance growth phenomenon doesn't occur if the displacement (or system) is at equilibrium to start (because multiplying the growth times zero). An example to keep in mind is the familiar swing. Once set in motion (with non-zero start), the swing motion is sustained by the appropriate (resonance matching) timing of swing motion with swing cycle, a parametric resonance. To capture the phenomenon, let's consider a 1-D spring system with mass and spring constant k:

$$\frac{d}{dt}(m\dot{x}) + kx = 0.$$

(Eqn. 3-63)

Let's rescale time to allow the presumed time-dependent m(t) to be separated:

$$d\tau = \frac{dt}{m(t)} \rightarrow \frac{d^2x}{d\tau^2} + mkx = 0.$$

Thus, without loss of generality (wlog), we can consider the problem in the form

$$\frac{d^2x}{dt^2} + \omega^2(t)x = 0,$$

(Eqn. 3-64)

that we could have arrived at from the start, allowing m=constant, but arriving at a form with a time-dependent system frequency $\omega(t)$.

Consider the case where $\omega(t)$ is periodic with frequency γ and period $T = 2\pi/\gamma$. If $\omega(t) = \omega(t + T)$, then the overall solution is invariant to $t \rightarrow t + T$. In turn, this means that the two independent solutions for displacements, $x_1(t)$ and $x_2(t)$ must also be invariant to $t \rightarrow t + T$, as can be seen by substitution in the above second order differential equation, aside from a non-time-dependent constant factor, thus the general solutions must satisfy:

$$x_1(t + T) = c_1 x_1(t) \text{ and } x_2(t + T) = c_2 x_2(t).$$

The most general solution is then:

$$x_1(t) = (c_1)^{t/T} P_1(t; T) \text{ and } x_2(t) = (c_2)^{t/T} P_2(t; T),$$

(Eqn. 3-65)

where $P_1(t; T)$ and $P_2(t; T)$ are purely periodic functions with period T. It turns out, however, that the constants c_1 and c_2 (that are exponentiated) in the solutions, have a relationship that forces one of them to always be the inverse of the other, thus there will always be an exponential growth term. Consider:

$$x_2(\ddot{x}_1 + \omega^2(t)x_1) = 0 \;\; and \;\; x_1(\ddot{x}_2 + \omega^2(t)x_2) = 0 \rightarrow \frac{d}{dt}(\dot{x}_1 x_2 - x_1 \dot{x}_2)$$
$$= 0$$

If $\dot{x}_1 x_2 - x_1 \dot{x}_2 = constant$, then with $t \rightarrow t + T$ the extra overall factor of $c_1 c_2$ that results must equal one, i.e., one c is the inverse of the other. This is referred to as parametric resonance but observe that it happens for any parametric driving frequency – practically speaking the accessible domain for this type of resonance is more restricted, as the derivation that follows relates. (Note: the boundary conditions might be such that the purely periodic functions are simply zero, a special case where exponential growth doesn't occur because it's zero to begin with.)

Since parametric resonance is a generic phenomenon when modulating a system parameter, is there an optimal frequency to do this? The answer is yes, and it's simply double the system's natural resonant frequency. In real-world applications with drag, this optimized driving frequency can often still operate at parametric (exponential growth) resonance. To show the specialized resonance in the drag-free case, start with the frequency parameter split into the time-independent resonant term $\omega_0{}^2$ and time-dependent offset multiplier term:

$$\omega^2(t) = \omega_0{}^2(1 + h\cos(\gamma t)),$$

(Eqn. 3-66)

where $h \ll 1$, and we choose $\gamma = 2\omega_0 + \epsilon$, where $\epsilon \ll \omega_0$. Let's try a solution of the form without parametric modulation, then account for that modulation by an offset to the natural frequency that matches with the parametric driver frequency:

$$x(t) = x_1(t) + x_2(t) = a(t)\cos\left(\left[\omega_0 + \frac{1}{2}\epsilon\right]t\right) + b(t)\sin\left(\left[\omega_0 + \frac{1}{2}\epsilon\right]t\right)$$

Substituting the above solution and expanding to first order in h, and first order in ϵ, where a(t) and b(t) vary slowly compared to ω_0, and assume $\dot{a}\sim\epsilon a$ and $\dot{b}\sim\epsilon b$ (later verified in result), first consider the trigonometric cross terms:

$$\cos\left(\left[\omega_0 + \frac{1}{2}\epsilon\right]t\right)\cos([2\omega_0 + \epsilon]t)$$
$$= \frac{1}{2}\cos\left(3\left[\omega_0 + \frac{1}{2}\epsilon\right]t\right) + \frac{1}{2}\cos\left(\left[\omega_0 + \frac{1}{2}\epsilon\right]t\right).$$

89

Note, the higher multiple frequency in the first term that results. Higher multiple frequency terms will contribute a higher order of smallness with respect to h, thus like higher order h, may be dropped in the first order analysis. The resulting equation is:

$$-(2\dot{a} + b\epsilon + \frac{1}{2}h\omega_0 b)\omega_0 \sin\left(\left[\omega_0 + \frac{1}{2}\epsilon\right]t\right) + (2\dot{b} - a\epsilon + \frac{1}{2}h\omega_0 a)\omega_0 \cos\left(\left[\omega_0 + \frac{1}{2}\epsilon\right]t\right) = 0$$

The coefficients of the trig terms must independently be zero. Let's try $a(t)\sim\exp(st)$ and $b(t)\sim\exp(st)$, which gives rise to the characteristic equations:

$$sa + \frac{1}{2}\left(\epsilon + \frac{1}{2}h\omega_0\right)b = 0 \text{ and } \frac{1}{2}\left(\epsilon - \frac{1}{2}h\omega_0\right)a - sb = 0 \rightarrow s^2$$
$$= \frac{1}{4}\left[\left(\frac{1}{2}h\omega_0\right)^2 - \epsilon^2\right].$$

Note that the solution range for exponential growth is where s is real, thus we have the constraint:

$$-\frac{1}{2}h\omega_0 < \epsilon < \frac{1}{2}h\omega_0.$$

3.8.7 Anharmonic Oscillations

Let's now consider a Lagrangian with terms at third order, but with a plan to work with expansions in the perturbation magnitude. In effect, we are solving differential equations using the classic method of successive approximations. What happens with this approach is the anharmonic oscillator is converted to a succession of driven harmonic oscillator problems. Let's start with a generic Lagrangian at third order:

$$L = \frac{1}{2}\sum_{\alpha}(\dot{\theta}_\alpha{}^2 - \omega_\alpha{}^2\theta_\alpha{}^2) + \sum_{\alpha,\beta,\gamma} C_{\alpha\beta\gamma}\dot{\theta}_\alpha\dot{\theta}_\beta\theta_\gamma - \sum_{\alpha,\beta,\gamma} D_{\alpha\beta\gamma}\theta_\alpha\theta_\beta\theta_\gamma$$

(Eqn. 3-67)

which leads to a second order E-L equation of the form:

$$\ddot{\theta}_\alpha + \omega_\alpha{}^2\theta_\alpha = f_\alpha(\theta_\alpha, \dot{\theta}_\alpha, \ddot{\theta}_\alpha).$$

(Eqn. 3-68)

This is then solved by the method of successive approximations, a perturbation analysis:

$$\theta_\alpha = \theta_\alpha^{(1)} + \theta_\alpha^{(2)}, \text{where } \theta_\alpha^{(2)} \ll \theta_\alpha^{(1)}, \text{and} \ddot{\theta}_\alpha^{(1)} + \omega_\alpha{}^2\theta_\alpha^{(1)} = 0.$$

90

This leaves the perturbation in terms of the effective force, but in the perturbation analysis we can approximate the generalized force's generalized coordinate dependency by the prior level of approximation, here:

$$\ddot{\theta}_\alpha^{(2)} + \omega_\alpha{}^2\theta_\alpha^{(2)} = f_\alpha\left(\theta_\alpha^{(1)}, \dot{\theta}_\alpha^{(1)}, \ddot{\theta}_\alpha^{(1)}\right).$$

(Eqn. 3-69)

At the second approximation we have the natural frequency of the system modified by various combination frequencies, such as $\omega_\alpha \pm \omega_\beta$, including $2\omega_\alpha$ and $\omega_\alpha = 0$. This process can be repeated, going to higher levels of approximation, but the fundamental frequencies ω_α in higher approximations are not equal to their unperturbed levels. To correct for this, modification is made such that periodic factors in the solution shall contain the exact frequencies. To be specific, let's consider the example of the following 1-D anharmonic oscillator [27]:

$$L = \frac{1}{2}m\dot{x}^2 - \frac{1}{2}m\omega_0^2 x^2 + xF(t),$$

$$where\ F(t) = -\frac{1}{3}max^2 - \frac{1}{4}m\beta x^3$$

(Eqn. 3-70)

for which we get:

$$\ddot{x} + \omega_0^2 x = -\alpha x^2 - \beta x^3.$$

(Eqn. 3-71)

Using the method of successive approximations described above (further details on this can be found in App. A), we have:

$$x = x^{(1)} + x^{(2)} + x^{(3)} + \cdots,$$

(Eqn. 3-72)

where we start with the homogeneous equation solution, i.e, where $x^{(1)} = a \cos \omega t$ with the exact value of ω where:

$$\omega = \omega_0 + \omega^{(1)} + \omega^{(2)} + \omega^{(3)} + \cdots,$$

(Eqn. 3-73)

and we get:

$$\frac{\omega_0^2}{\omega^2}\ddot{x} + \omega_0^2 x = -\alpha x^2 - \beta x^3 - \left(1 - \frac{\omega_0^2}{\omega^2}\right)\ddot{x}.$$

(Eqn. 3-74)

To go to the next level of approximation, let's consider $x = x^{(1)} + x^{(2)}$ and $\omega = \omega_0 + \omega^{(1)}$, and omitting terms above second order of smallness:

91

$$\ddot{x}^{(2)} + \omega_0^2 x^{(2)} = -\alpha a^2 \cos^2 \omega t + 2\omega_0 \omega^{(1)} a \cos \omega t$$

(Eqn. 3-75)

now choose $\omega^{(1)} = 0$ to arrive at a simple solution (we choose the ω modifications at successive approximations for similar decoupling or simplification):

$$x^{(2)} = -\frac{\alpha a^2}{2\omega_0^2} + \frac{\alpha a^2}{6\omega_0^2} \cos 2\omega t$$

(Eqn. 3-76)

Going to the next level of approximation with $x = x^{(1)} + x^{(2)} + x^{(3)}$ and $\omega = \omega_0 + \omega^{(2)}$, we get:

$$\ddot{x}^{(3)} + \omega_0^2 x^{(3)} = -2\alpha x^{(1)} x^{(2)} - \beta \left(x^{(1)}\right)^3 + 2\omega_0 \omega^{(2)} x^{(1)}$$

(Eqn. 3-77)

$$\ddot{x}^{(3)} + \omega_0^2 x^{(3)} = a^3 \left[\frac{\beta}{4} - \frac{\alpha^2}{6\omega_0^2}\right] \cos 3\omega t$$
$$+ a \left[2\omega_0 \omega^{(2)} + \frac{5\alpha^2 a^2}{6\omega_0^2} - \frac{3}{4}a^2\beta\right] \cos \omega t$$

(Eqn. 3-78)

where, again, we choose $\omega^{(2)}$ such that the term on the right is zero for a simple solution:

$$\omega^{(2)} = -\frac{5a^2\alpha^2}{12\omega_0^3} + \frac{3\beta a^2}{8\omega_0}$$

(Eqn. 3-79)

and,

$$x^{(3)} = \frac{a^3}{16\omega_0^2} \left[\frac{\alpha^2}{3\omega_0^2} - \frac{\beta}{2}\right] \cos 3\omega t.$$

(Eqn. 3-80)

Parametric resonance is mainly evident in studies of systems acting under small oscillations and involves time variation of the system parameters – such as the point of support for a pendulum (to be described in the next section). Forced oscillations, with or without damping, have a dispersion-type frequency dependence on the absorption of energy from the driver. There is resonance at the natural frequency of the system. For motions that have been substantially excited we get into the nonlinear regime of the kinetic and potential energy terms in the Lagrangian. Anharmonic, or non-linear, oscillations (like in the previous section) get mixing due to the non-linearities which result in combination frequencies that themselves can appear resonant. In this regard, the method of successive approximations must be used carefully, in a manner consistent with not having self-resonant terms via the mixing.

3.8.8 Motion in rapidly oscillating field (a.k.a. two-timing analysis)

Consider motion in a potential U with period T where a rapidly oscillating force is applied,

$$m\ddot{x} = -\frac{dU}{dx} + f, \quad \text{where } f = f_1 \cos \omega t + f_2 \sin \omega t,$$

$$\text{and where } \omega \gg \frac{1}{T}$$

(Eqn. 3-81)

We do not assume that $f \ll U$ or even $f < U$, rather we assume an outcome with small oscillations on top of the smooth path the particle would traverse if only under the potential U:

$$x(t) = X(t) + \varepsilon(t), \text{where } \overline{\varepsilon(t)} = 0.$$

(Eqn. 3-82)

This is sometimes referred to as a two-timing analysis [30]. Substituting, we then get to first order in Taylor expansions:

$$m\ddot{X} + m\ddot{\varepsilon} = -\frac{dU}{dx} - \varepsilon \frac{d^2U}{dx^2} + f(X,t) + \varepsilon \frac{\partial f}{\partial X}.$$

(Eqn. 3-83)

Now all of the first order terms in ε are negligible compared to the other terms, except for the $\ddot{\varepsilon}$ term, since the frequency factors are assumed very large (since rapidly oscillating). Splitting the smooth trajectory ($X(t)$ trajectory with $f = 0$) and the rapidly oscillating part, we get for the latter:

$$m\ddot{\varepsilon} = f(X,t) \rightarrow \varepsilon = -\frac{f}{m\omega^2}$$

(Eqn. 3-84)

Now consider the average with respect to time on the first-order equation, all stand-alone first-powers of ε and f will be zero:

$$m\ddot{X} = -\frac{dU}{dx} + \varepsilon \overline{\frac{\partial f}{\partial X}} = -\frac{dU}{dx} - \frac{1}{m\omega^2} \overline{f \frac{\partial f}{\partial X}} = -\frac{dU_{eff}}{dx},$$

where,

$$U_{eff} = U + \frac{\overline{f^2}}{2m\omega^2}, \quad \text{also have } U_{eff} = U + \frac{(f_1^2 + f_2^2)}{4m\omega^2} = U + \frac{1}{2}m\overline{\varepsilon^2}$$

(Eqn. 3-85)

To see how this is exhibited in practice, consider the pendulum whose point of support is undergoing rapid *horizontal oscillations*:

$x = l \sin\varphi + a \cos\gamma t$ and $\dot{x} = l\dot\varphi \cos\varphi - a\gamma \sin\gamma t$

$y = l \cos\varphi$ and $\dot{y} = -l\dot\varphi \sin\varphi$

$U = -mgl \cos\varphi$

$$L = T - U = \frac{1}{2}m(l\dot\varphi)^2 - ml\dot\varphi a\gamma \cos\varphi \sin\gamma t + mgl \cos\varphi$$

making use of the freedom to add a total time derivative, $\frac{d}{dt}(mla\gamma \sin\varphi \sin\gamma t)$, to get:

$$L = T - U = \frac{1}{2}m(l\dot\varphi)^2 + mla\gamma^2 \sin\varphi \cos\gamma t + mgl \cos\varphi$$

Using the Euler-Lagrange equation we then get:

$$ml^2\ddot\varphi = mla\gamma^2 \cos\varphi \cos\gamma t - mgl \sin\varphi = -\frac{dU}{dx} + f_\varphi,$$

where,

$$f_\varphi = mla\gamma^2 \cos\varphi \cos\gamma t$$

Using the relation from the prior discussion:

$$U_{eff} = U + \frac{\overline{f_\varphi^2}}{2m\gamma^2} = mgl\left[-\cos\varphi + \frac{a^2\gamma^2}{4gl}\cos^2\varphi\right].$$

Solving for $\frac{dU_{eff}}{d\varphi} = 0$ we get solutions at $\sin\varphi = 0$ and $\cos\varphi = 2gl/a^2\gamma^2$, where the existence of the latter solution requires $2gl < a^2\gamma^2$.

Similarly, we could consider the pendulum whose point of support is undergoing rapid *vertical oscillations*:

$x = l \sin\varphi$ and $\dot{x} = l\dot\varphi \cos\varphi$

$y = l \cos\varphi + a \cos\gamma t$ and $\dot{y} = -l\dot\varphi \sin\varphi - a\gamma \sin\gamma t$

$U = -mgl \cos\varphi + mga \cos\gamma t$

$$L = T - U = \frac{1}{2}m(l\dot\varphi)^2 + ml\dot\varphi a\gamma \sin\varphi \sin\gamma t + \frac{1}{2}ma^2\gamma^2 \sin^2\gamma t$$
$$+ mgl \cos\varphi - mga \cos\gamma t$$

Dropping pure time-dependent functions and making use of the freedom to add a total time derivative, $\frac{d}{dt}(mla\gamma \cos\varphi \sin\gamma t)$, to get:

$$L = T - U = \frac{1}{2}m(l\dot\varphi)^2 + mla\gamma^2 \cos\varphi \cos\gamma t + mgl \cos\varphi$$

Using the Euler-Lagrange equation we then get:

94

$$ml^2\ddot{\varphi} = -mla\gamma^2 \sin\varphi \cos\gamma t - mgl\sin\varphi = -\frac{dU}{dx} + f_\varphi,$$

where,

$$f_\varphi = -mla\gamma^2 \sin\varphi \cos\gamma t$$

Using the relation from the prior discussion again:

$$U_{eff} = U + \frac{\overline{f_\varphi}^2}{2m\gamma^2} = mgl\left[-\cos\varphi + \frac{a^2\gamma^2}{4gl}\sin^2\varphi\right].$$

Solving for $\frac{dU_{eff}}{d\varphi} = 0$ we get solutions at $\varphi = 0$ and $\varphi = \pi$, where the existence of the latter solution requires $2gl < a^2\gamma^2$.

Chapter 4. Classical Measurement

4.1 Capturing small measurements in time-integrable systems

Measurement with the highest sensitivity occurs where the measurement event is repeated, often in arrangements where a key value is summed over time. Thus, it is natural to look to time-integrable systems as a key component of a sensitive detector. An oscillator is an example of such a system, for which a brief recap is provided next. After that we do one last generalization, the addition of noise fluctuations (fundamentally present due to thermal noise sources) to get a description of actual experimental limits. Initially, to build from the CM results shown in Ch. 3 we will develop the damped driven oscillator with noise and see what minimal detectable force acting on the oscillator (mass) is possible. This describes a "contact" method for force detection.

Direct contact methods for actual detection are more typically based on strain gauges or piezoelectric elements that can directly couple into electrical (resonance) circuits (note conversion of signal to electronic form, which will be the norm). Indirect contact methods based on capacitance meters do best in this category, where the measurement of a displacement directly alters the capacitance (via plate separation directly related to displacement). The resting capacitance is chosen in a circuit operating at resonance (or on the steep part of the resonance curve) [51] such that circuit frequency shifts are most notable by a secondary circuit (indirect contact) measurement device. Examples of the capacitance meters get into circuit descriptions that, although straightforward [52], are outside the scope of this description so won't be discussed further.

Optical non-contact methods offer the greatest sensitivity, and those will be briefly discussed following more explicit results for the contact methods (since the presentation of an oscillator direct-contact detector demonstrates many of the key concepts and limiting factors). Note that the most extreme "non-contact" detection is quantum non-demolition, but that will not be discussed. Notes from the LIGO project, and were obtained from Prof. Drever's course Ph118 ca. 1988 (in Appendix B, the ~1988 the LIGO contact list shows less than 30 on the project, including myself a graduate student at the time, there are now over 3000 contributors on this project worldwide).

4.1.1 Recap of damped driven oscillator
For the damped driven oscillator we have the ODE:

$$\ddot{x} + 2\lambda\dot{x} + \omega^2 x = \left(\frac{F}{m}\right)\exp i\gamma t,$$

(Eqn. 4-1)

with solution:

$$x(t) = a\exp(-\lambda t)\cos(\omega t + \alpha) + b\cos(\gamma t + \delta) \cong b\cos(\gamma t + \delta),$$

(Eqn. 4-2)

where

$$b = \frac{F}{m\sqrt{(\omega^2 - \gamma^2)^2 + (2\lambda\gamma)^2}} \quad and \quad \tan\delta = \frac{(2\lambda\gamma)}{(\omega^2 - \gamma^2)}.$$

(Eqn. 4-3)

Once steady state motion is achieved, $x(t) \cong b\cos(\gamma t + \delta)$, energy absorbed from the external force matches that dissipated in friction. We have the rate of dissipation due to friction previously as $-2\mathcal{F}$, where $\mathcal{F} = \frac{1}{2}\alpha\dot{x}^2 = \lambda m b^2 \gamma^2 \sin^2(\gamma t + \delta)$, with time average: $2\bar{\mathcal{F}} = \lambda m b^2 \gamma^2$. Thus the energy absorbed per unit time is $\lambda m b^2 \gamma^2$. Now if we want the integral of the energy absorbed at all driving frequencies, the absorption will be dominated by the frequencies near resonance, for which the integral approximates as $\pi F^2/4m$.

4.1.2 Damped driven oscillator with noise fluctuations
Let's now consider the damped driven oscillator with noise fluctuations and determine the minimum detectable force that the system can provide. This is the scenario, with realistic noise fluctuations, that provides an accurate limit to the sensitivity of measurement. Let's start with the new ODE with added noise fluctuations term F_{fl}:

$$\ddot{x} + 2\lambda\dot{x} + \omega^2 x = F(t) + F_{fl},$$

(Eqn. 4-4)

where the steady state result from before, without fluctuation noise forces, was $x(t) \cong b\cos(\gamma t + \delta)$. Is there still a steady state but with a slightly more general form? First consider that the amplitude relation time is given by $\tau_m = 1/\lambda$ and we are assuming that the intention is to do precise measurements, so we seek a minimal damping, thus a maximal relaxation time τ_m, thus effectively steady state compared to the time of measurement and the time of the $F(t)$ effect meant to be detected. We thus will have the steady state form indicated with possible time dependence in the constants at a guess. Trying the guess and validating it then proves this to be correct [53] and [54]. Switching now to the notation of Braginsky [51], we will summarize the derivation of Braginsky shown in the Appendix of [51] titled "Statistical Criteria for the Determination of the Excitation of an oscillator by an external force":

$$x(\tau) \cong A(\tau) \sin\big(\omega_0 \tau + \varphi(\tau)\big) \quad and \quad \overline{A(\tau)} \gg \frac{1}{\omega_0} \frac{dA(\tau)}{d\tau}.$$

(Eqn. 4-5)

Our assertion of a detection event will be a probabilistic one, especially given the addition of a stochastic process (noise fluctuations). We wish to consider the probability that a force event $F(t)$ occurring in time $\hat{t}$ that falls within the timeframe of the measurement. The detectability of such an event requires discerning it from false signals from the fluctuation noise F_{fl}. In turn, the nature of the detectability must be examined for both. In both cases what we are looking for is a change in the amplitude of oscillation according to the difference $A(\tau) - A(0)$, and in the case of the fluctuation noise this limit must be qualified to be valid with probability "$1 - \alpha$". This approach is motivated by the expression from [54] for the probability density of an arbitrary distribution of amplitudes of oscillation after event time $\hat{t}$:

$$P[A(\hat{t})|A(0)]$$
$$= \frac{A(\hat{t})}{\sigma^2(1 - \varepsilon^2)} I_0 \left(\frac{\varepsilon A(0)A(\hat{t})}{\sigma^2(1 - \varepsilon^2)} \right) \exp\left(-\frac{\big(A(\hat{t})\big)^2 + \varepsilon\big(A(0)\big)^2}{2\sigma^2(1 - \varepsilon^2)} \right),$$

(Eqn. 4-6)

where,

$$\varepsilon = e^{(-\hat{t}/\tau_m)} \quad and \quad \sigma^2 = \overline{A(\tau)^2}.$$

The statistical error of the first kind formalism (with "$1 - \alpha$") now takes the form:

$$1 - \alpha = \int_{A(0)}^{A(\hat{t})} P[A(\hat{t})|A(0)] dA(\hat{t}).$$

(Eqn. 4-7)

Following Braginsky's analysis, we will now consider solving the integral for two cases: $A(0) = 0$ and $A(0) = \sigma$. We will find that the evaluation of minimal detectable force is roughly the same regardless of initial value of the amplitude, while the energy exchange with the oscillator is significantly affected by initial amplitude. Also, following Braginsky, we shall assume our noise source is purely a thermal noise source. This is the best case scenario as thermal noise sources are fundamental in physical systems in a variety of ways (see [24] for derivation of these noise sources in circuits, for example). If we assume "just" thermal noise we then have, according to the thermalization temperature T the following:

$$\sigma^2 = \frac{k_B T}{k}, \quad where \ \omega_0 = \sqrt{k/m}.$$

(Eqn. 4-8)

Solving the integral and substituting we then get:

$$[A(\hat{t})]_{1-\alpha} = 2\sigma\sqrt{(\hat{t}/\tau_m)\ln(1/\alpha)}.$$

(Eqn. 4-9)

Thus, if we start a detection event with $A(0) \cong 0$, and we see the amplitude grow in time $\hat{t}$ such that $A(\hat{t}) > [A(\hat{t})]_{1-\alpha}$, then we have with probability, or "reliability", $(1 - \alpha)$, that an event has occurred. As noted by Braginsky, what we have is only a threshold condition thus far describing what to do if the threshold is met. If the threshold is met, then we are saying no detection event, e.g., that $F(t) = 0$, but this may only be due to an unfortunate cancellation of event force and fluctuation forces. To assess the error that can be introduced from this, Braginsky introduces a measurement of a statistical error of the second kind corresponding to the probability of having $F(t) \neq 0$ while still having the below threshold event $A(\hat{t}) < [A(\hat{t})]_{1-\alpha}$. Specifically, consider the force $F(t)$ when there is no fluctuation force present and such that the change in the amplitude in time $\hat{t}$ is to a value Γ that is greater than the threshold, such that we have

$$\gamma = \Gamma/[A(\hat{t}) - A(0)]_{1-\alpha}$$

(Eqn. 4-10)

with $\gamma \geq 1$. This lays the foundation for evaluating the error of the second kind (further details are in [51]). The conclusion is that a simple constant factor, ~ 1, is all that would modify the threshold condition for detection event.

Let's now relate the minimal detectable change in amplitude to the energy imparted or extracted from the oscillator using the form with γ above:

$$\Delta E = k\gamma^2[A(\hat{t})]_{1-\alpha}^2 = 2\ln(1/\alpha)\,(2\hat{t}/\tau_m)\gamma^2 k_B T.$$

(Eqn. 4-11)

Returning to the simple case of $F(t) = F_0 \sin(\omega t)$ in time interval from 0 to $\hat{t}$ (and zero force outside that time interval), then we have the linear growth in amplitude according to:

$$\Gamma = \frac{F_0 \hat{t}}{2m\omega}, \quad where \ \omega = \sqrt{k/m}$$

and requiring that $\Gamma > [A(\hat{t}) - A(0)]_{1-\alpha}$ then gives the minimum detectable F_0:

$$[F_0]_{min} = \rho\sqrt{4k_BTm/(\hat{t}\tau_m)},$$

(Eqn. 4-13)

where ρ is a dimensionless reliability factor that ranges between 2.45 and 4.29 for typical reliability values α (see Table A1 in [51])., A similar analysis for the case where $A(0) \cong \sigma$ at the start of the detection event reduces to the same formula with reliability factors ranging between 1.96 and 3.88. Thus, the minimal detectable force is roughly the same regardless of initial value of the amplitude and has the form:

$$[F_0]_{min} \propto \sqrt{\frac{4k_BTm}{(\hat{t}\tau_m)}}.$$

(Eqn. 4-14)

4.1.3 Optical non-contact methods

There are two types of optical measurement that we will focus on here: (i) knife edge; and (ii) self interference. The knife edge methods involve an optical lever in some capacity. If we shine a laser beam on a mirror and measure its fluctuations on a screen distance D away, then the projected signal is twice as great if we simply double the projection distance to 2D. More common, and a blend of type (i) and (ii), is to use a diffraction grating, where the gain effect is multiplied according to the separation in the movable diffraction grating that is part of a beam-transmission measurement (involving a second, fixed, diffraction grating). The most sensitive of the optical self-interference type of detection events, however, typically involves a Michelson-Morley interferometer. The basic idea is that the beam is split and allowed to interfere with itself such that perfect cancellation is tuned at the transmitted part of the beam splitter. When a displacement in the mirror (or mirror-cavity distance) occurs, we then see a slip from the cancelled state, and see a flash of light according to the extent of non-cancellation, which is related to the strength of the signal. As with many of the detection methods, an evaluation of sensitivity often looks promising but, actually, obtaining the physical device parameters needed are often unobtainable. With the interferometric approaches, however, what is needed is often within reach, using very powerful lasers, highly reflective mirrors, exquisitely stabilized mirrors and beam-splitter mirror, for starters. It turns out this can be done, but it's a matter of scale.

Work I participated on involving the prototype LIGO detector in the 1980's was an instance where the interferometric methods were demonstrated to work extremely well. But the prototype interferometer arms were 20m long, not 2km, as they would eventually need to be. So the scale of vacuum was very different (the laser interferometer cavities are maintained at high vacuum to eliminate noise, and more importantly, avoid a destructive process on the (very expensive) highly reflective mirrors (an EM effect to be discussed in [40], results in uncharged "dust" taking an effective charge and, in the non-uniform electric field of the cavity, the result is the dust is driven into the mirrors causing their steady degradation). This, and other scaling issues required another 30 years of development until the LIGO project finally came online with the first gravitational wave observatory (Nobel Prize for Kip Thorne,et al.). In the 1980's when I participated for a few years (before shifting into more theoretical issues, to be described in [45,46]) the LIGO group was quite small (about 30, see the old Directory in Fig. B.1). The scaling by 100x in the device size was partly met by a 100x rescaling in the group effort by the year 2020.

A proper description of the LIGO detection methodology would take us to far afield into laser noise properties and optical cavity properties, but a high level description is still given. First, the "L-shaped" interferometer is doubly important for the type of detection event sought, which for LIGO was a gravitational wave. Such a wave would be measureable only via its quadrupole effect (with orthogonal detector arms, see Book 3 for details) whereby one arm of the interferometer is lengthened while the other is shortened, providing a change in the interference signal (this for the quadrupole wave hitting the detector perfectly transversely and aligned on the detector arms). Second, the laser noise (multimodality) directly relates to shifts in the main mode that is being "locked on," which is a noise problem, thus requires something to "clean" the laser noise. At the time I was working at LIGO the resonant cavity that was employed for this task was named by Ron Drever as the "dewiggler". Thus, there is a laser cavity (high powered) that feeds into a mode cleaner (the dewiggler) that then feeds into the "L-shaped" interferometer. And, third, there is the matter of stabilizing the arm lengths against positional fluctuations in the frequency band of interest for detection. In essence the end-mirrors and beam splitter mirror must all be servo'd to fixed position relative to each other (the whole system floats with respect to the surrounding vacuum chamber while relatively 'locked'). In the end, specialized signal processing is needed for detection of a known signal profile (or group of profiles). In essence, a specialized filter based on matching to the signal sought is employed for optimal detection capability.

4.2 Measurement Theory – Random Variables and Processes

Many experiments are described where there is a predicted frequency or other measureable characteristic. We would like to get an "accurate measurement", but what does this mean? To begin, consider a set of measurements for some circumstance, perhaps as simple as the repeated measurement of something. In measurement theory the set of such measurements, in the simplest non time-varying cases, is viewed as a sample from a single type of background distribution. By doing repeated measurements (x_N) we know intuitively we get a better, or 'safer', measurement, but why is this? It turns out it is simple to derive the property that the sample variance decreases with number of measurements taken. How many measurements to take then turns into how tight you want your "error bars" (the region delineated from one standard deviation, or σ (sigma), below the mean to one standard deviation above). We will see that $Var(\bar{x}_N) = \sigma^2/N$, where σ is the standard deviation of a single measurement of the random variable (X), and Var is the variance (std. dev. squared) of the repeated measurement. This computation is known as computing the sigma of the mean and we get that $\sigma_\mu = \sigma/\sqrt{N}$, thus we can improve our measurement accuracy (reduced sigma on the mean) according to the number of measurements taken (N). The above core result (justification for repeated measurements in the experimental process) as well as others will now be outlined in further detail. A number of technical terms have already arisen in the discussion above, however, thus a brief review of core terminology and definitions will now be given first.

Definitions
Most of the definitions that follow in this section are detailed further in [55].

Random Variable
A Random Variable X is an assignment of a number, $x(\theta)$, to every outcome θ of X.

Stochastic Process
A Stochastic Process is an assignment of time-parameter dependent number, $x(\theta,t)$, to every outcome θ of X.

Viewed as an index, if the time-parameter t is continuous, then we have a continuous-time process, otherwise it is a discrete-time process. Let's work with discrete-time processes for now and provide more definitions - - laying foundation for repeated experimental measurements scenario:

The Expectation, E(X), of random variable X

The expectation, E(X), of random variable X is defined to be:

$$E(X) \equiv \sum_{i=1}^{L} x_i \, p(x_i) \; if \; x_i \in \mathcal{R}.$$

(Eqn. 4-15)

Similarly, the expectation, E(g(X)), of a function g(X) of random variable X is:

$$E(g(X)) \equiv \sum_{i=1}^{L} g(x_i) \, p(x_i) \; if \; x_i \in \mathcal{R}.$$

Now consider the special case where $g(x_i) = -log(\,p(x_i)\,)$, which gives rise to Shannon's entropy:

$$H(X) \equiv E[g(X)] = - \sum_{i=1}^{L} p(x_i) \, log(p(x_i)) \; if \; \; p(x_i) \in \mathcal{R}^{+},$$

For Mutual Information, similarly, use $g(X,Y) = log(p(x_i, y_i)/p(x_i)p(y_i))$ to get:

$$I(X;Y) \equiv E[g(X,Y)] = \sum_{i=1}^{L} p(x_i, y_i) \, log(p(x_i, y_i)/p(x_i)p(y_i)) ,$$

and if $p(x_i)$, $p(y_i)$, $p(x_i, y_i)$ are all $\in \mathcal{R}^{+}$, then this is equivalent to the Relative Entropy between a joint distribution and the same distribution if the random variables are independent, a.k.a., it is the Kullback-Leibler Divergence: $D(\,p(x_i, y_i) \,||\, p(x_i)p(y_i)\,)$ that is prevalent in information theory [24].

Jensen's Inequality

The foundation has been laid for a simple proof of Jensen's inequality, which is provided next. This inequality is a key maneuver employed in other definitions to follow (Hoeffding).

Let $\varphi(\cdot)$ be a convex function on a convex subset of the real line: φ: $\chi \rightarrow \mathcal{R}$. Convexity by definition: $\varphi(\lambda_1 x_1 + ... y_n x_n) \leq \lambda_1 \varphi(x_1) + \; ... \; + \lambda_n \varphi(x_n)$, where $\lambda_i \geq 0$ and $\Sigma \, \lambda_i = 1$. Thus, if $\lambda_1 = p(x_1)$, we satisfy the relations for line interpolation as well as discrete probability distributions, so can rewrite in terms of the Expectation definition:

$$\varphi(\,E(X)\,) \leq E(\,\varphi(X)\,).$$

Let's apply this to get a relation involving Shannon Entropy by choosing $\varphi(x) = -log(x)$, which is a convex function, therefore we have that:

$$log(\,E(X)\,) \geq E(\,log(X)\,) = -H(X).$$

Variance

$$Var(X) \equiv E(\,[X - E(X)]^2\,) = \sum_{i=1}^{L}(x_i - E(X))^2 p(x_i) = E(X^2) - (E(X))^2$$

(Eqn. 4-16)

Sample Variance

$$Var_N(X) = \frac{1}{N-1}\sum(x_i - E(x))^2$$

(Eqn. 4-17)

Chebyshev's Inequality

For $k>0$, $P(|X - E(X)|>k) \le Var(X)/k^2$

(Eqn. 4-18)

Proof:
$$\begin{aligned}
Var(X) &= \sum_{i=1}^{L}(x_i - E(X))^2 p(x_i) \\
&= \sum_{\{x_i|\,|x_i - E(X)|>k\}}(x_i - E(X))^2 p(x_i) \\
&\quad + \sum_{\{x_i|\,|x_i - E(X)|\le k\}}(x_i - E(X))^2 p(x_i) \\
&\ge k^2\, P(|X - E(X)|>k)
\end{aligned}$$

Repeated measurement and the sigma of the mean

Let X_k be independent identically distributed (iid) copies of X, and let X be the real number "alphabet". Let $\mu = E(X)$, $\sigma^2 = Var(X)$, and denote

$$\bar{x}_N = \frac{1}{N}\sum_{k=1}^{N} X_k$$
$$E(\bar{x}_N) = \mu$$
$$Var(\bar{x}_N) = \frac{1}{N^2}\sum_{k=1}^{N} Var(X_k) = \frac{1}{N}\,\sigma^2$$

Thus, for repeated measurements, the sigma of the mean is $\sigma_\mu = \sigma/\sqrt{N}$, as mentioned previously. Note that if we continue the analysis of this scenario we get for the Chebyshev relation:

$$P(|\bar{x}_N - \mu|>k) \le Var(\bar{x}_N)/k^2 = \frac{1}{Nk^2}\,\sigma^2.$$

(Eqn. 4-19)

from which the Law of Large Numbers can be derived.

The Law of Large Numbers, Weak Form (Weak-LLN)

The LLN will now be derived in the classic "weak" form. (The "strong" form is derived in the modern mathematical context of Martingales in a later section.) As $N \to \infty$ we get what is known as the Law of Large Numbers (weak), where $P(|\bar{x}_N - \mu|>k) \to 0$, for any $k>0$. Thus, the arithmetic mean of a sequence of iid r.v.s converges to their common expectation. The weak form has convergence "in probability", while the strong form will have convergence "with probability one".

4.3 Collisions & Scattering

Let's now turn to consideration of collision and scattering. This is an application of the Lagrangian analysis that is usually straightforward, especially when considering classical scattering for which there is always an answer [56]. We will do this in the Lagrangian-based formulation, with energy as a conserved quantity, and consider the unbounded trajectories (incoming and outgoing). A very brief, but formal description of classical scattering along the lines of Reed&Simon [56] will be given afterwards, which can then directly transition to a quantum scattering description (as shown in [56]). Before embarking on the formal description, let's first get the basics down by a re-examination of Rutherford scattering (1911) [57] and Compton scattering (1923) [73], the former moving us from the plum-pudding model of the atom to the modern with compact nucleus and electron cloud, and revealing the central role of alpha; the latter providing direct evidence of 4-vector mathematics (evidence of SR). (Had Compton scattering been observed before 1905, it would have been another part of physics, accessible from the classical experimental devices of the time, indicating SR.)

The focus of the classical mechanics thus far has been on the mathematical theory and not on he observed parameters of the elementary particles observed or the phenomenological description of "ponderable media" (to be discussed, for the classical mechanical setting, in Sec. 5.1 for Rigid Bodies and Sec. 5.2 for Material Bodies). And this has been done to clearly separate the fundamental particle parameters and phenomenological parameters from the mathematical structure, including from fundamental mathematical parameters. In Sec. 4.3 on Scattering and Ch. 5 on Collective Motion (an early exploration of Material properties), however, the physical parameters are unavoidable and also pertain to key experiments demonstrating the strength of certain experimental models, so they will begin to appear in the presentation. We start with Rutherford scattering [57], which is simply Coulomb scattering at low speed (non-relativistic). We get a formula, and it fits remarkably well to experiment if we assume the modern atomic model (positive, compact nucleus, with negative electron cloud). There is only one "fit parameter" in the formula and it is the dimensionless parameter alpha. Thus, we have our first appearance of alpha in the classical mechanics discussion (grouped as $\alpha\hbar$), and it directly relates to atomic properties (charge), electromagnetic properties (permittivity of free space), special relativistic properties (speed of light) and quantum properties (Planck's constant). (Note, alpha had already appeared in early QM efforts, as the fine-structure constant, in spectrographic analysis by Sommerfeld [58], as will be discussed in Book 4.) Before working thru several examples, the Compton Scattering is shown also. The Compton scattering experiment was actually done and the description draws upon Caltech Ph 7 lab notes where the Compton experiment was done as part of a standard lab requirement for Physics undergraduates. Use of coincidence detection capability allows for acquisition of excellent data. The validation of Compton's scattering formula, in turn, serves to demonstrate:(i) that light cannot be explained purely as a wave phenomenon (further quantum discussion delayed until Book 4 [42]); and (ii) that consistency requires use of the relativistic energy-momentum 4-vector relation (SR covered in Book 2 [40]).

In scattering we often seek to examine the amount of scattering (or probability of scattering) into a given angle (such as with Rutherford). The measure of the probability of a given process is thereby reduced to evaluation of the relevant "cross section." Further details on these definitions and conventions will be brought out in the course of examining Rutherford scattering discussed next.

4.3.1. Rutherford Scattering

Consider two charged point particles interacting under a central Coulomb potential. The classical central potential allows decoupling of center of mass motion and relative motion, we thus choose a convenient "frame" with particle 1 in motion (incident on particle 2) with parameters: m_1, $q_1 = Z_1 e$ (where e is the fundamental charge, and Z_1 is a positive integer), and a non-zero velocity v_1 measured when very far away.

Sec. 3.7 describes motion in a central Coulomb field (with two point particles having opposite charges), for which we obtained the solution:
$$p = r(1 + e \cos \theta).$$

(Eqn. 4-20)

The general solution (including unbounded motion) is closely related and is given by:
$$u = u_0 \cos(\theta - \theta_0) - C, where \ u = \frac{1}{r}.$$

(Eqn. 4-21)

If we now consider the boundary conditions, asymptotically, for the ingoing/outgoing scattering of interest, we must have solutions that satisfy:
$$u \to 0 \ and \ r \sin \theta \to b \ as \ \theta \to \pi,$$

where b is the impact parameter. When solved to provide a relation between b and the deflection angle we get:

$$b = \frac{Z_1 Z_2 e^2}{4\pi\epsilon_0 m v_1^2} \cot\frac{\theta}{2}.$$

(Eqn. 4-22)

We've now obtained a relation $b(\theta)$ from which the cross-section is easily obtained using the standard formula:

$$\frac{d\sigma}{d\Omega} = \frac{b}{\sin\theta}\left|\frac{db}{d\theta}\right|.$$

(Eqn. 4-23)

Before moving on, however, let's re-derive this formula and in doing so know precisely what is meant by the "scattering cross section". The formal definition is:

$$\frac{d\sigma}{d\Omega}d\Omega = \frac{number\ scattered\ into\ d\Omega\ per\ unit\ time}{incident\ intensity}.$$

(Eqn. 4-24)

Consider an ingoing (axial) beam of particles, with intensity uniform, with impact parameter between b and $b + db$, the number of particles incident with desired impact parameter are then:

$$2\pi I b|db| = I\frac{d\sigma}{d\Omega}d\Omega,$$

(Eqn. 4-25)

where use is made of the definition of the number of particles scattered into solid angle $d\Omega$. Since the scattering potential is radially symmetric we have $d\Omega = 2\pi\sin\theta\,d\theta$, thus:

$$\frac{d\sigma}{d\Omega} = \frac{b}{\sin\theta}\left|\frac{db}{d\theta}\right|.$$

Applying the formula:

$$\frac{d\sigma}{d\Omega} = \left(\frac{Z_1 Z_2 e^2}{8\pi\epsilon_0 m v_1^2 \sin^2\frac{\theta}{2}}\right)^2 = \left(\frac{Z_1 Z_2 (\alpha\hbar c)}{2m v_1^2 \sin^2\frac{\theta}{2}}\right)^2, \quad where\ \ \alpha = \frac{e^2}{4\pi\epsilon_0\hbar c}.$$

(Eqn. 4-26)

4.3.2. Compton Scattering

Let's consider X-ray scattering next. Not only are X-rays scattered into various angles in a particle-like manner, the 'particle' itself appears to change in that the X-ray wavelength shifts according to the amount (angle) of scattering. Compton will consider photons in a particle-wave formalism by using the formula of Einstein's photovoltaic effect. Compton will also consider the photons in a relativistic setting, such that

108

the special relativity energy-momentum is the representation of the total energy. The scattering experiment will consist of an incoming (collimated) X-ray beam striking a fixed electron with scattering of X-ray and recoil of he electron. Thus we have from conservation on energy (relativistic):

$$hf + mc^2 = hf' + \sqrt{(pc)^2 + (mc^2)^2},$$

(Eqn. 4-27)

where f is the frequency of the incoming X-ray (using Einstein relation with Planck's constant h), m is the (rest) mass of the electron, c is the speed of light, mc^2 is thus the rest energy of the electron according to Einstein's special relativity. On the RHS, we have the new X-ray frequency f', the non-zero electron recoil momentum p, such that the recoil electron's relativistic energy-momentum is $\sqrt{(pc)^2 + (mc^2)^2}$. For conservation of 4-momentum we have:

$$p = p_\gamma - p_{\gamma'}$$

(Eqn. 4-28)

which can be rewritten as:

$$(pc)^2 = \left(p_\gamma c\right)^2 + \left(p_{\gamma'}c\right)^2 - 2\left(p_\gamma c\right)\left(p_{\gamma'}c\right)\cos\theta,$$

(Eqn. 4-29)

and when combined with the conservation of energy relation we get the famous Compton equation:

$$\frac{c}{f'} - \frac{c}{f} = \frac{h}{mc}(1 - \cos\theta).$$

(Eqn. 4-30)

The angular distribution on the scattered photons is described by the Klein-Nishina formula:

$$\frac{d\sigma}{d\Omega} = \frac{\left(\frac{1}{2r_0}\right)[1 + \cos^2\theta]}{\left[1 + 2\varepsilon\sin^2\left(\frac{\theta}{2}\right)\right]}\left\{1 + \frac{4\varepsilon^2\sin^4\left(\frac{\theta}{2}\right)}{[1 + \cos^2\theta]\left[1 + 2\varepsilon\sin^2\left(\frac{\theta}{2}\right)\right]}\right\}$$

(Eqn. 4-31)

Exercise. Derive the Klein-Nishina formula.

4.3.3. Theoretical Discussion and Examples
Thus far the scattering descriptions have involved potentials with attractive forces, like gravity or Coulomb with opposite charges. They could also involve repulsive forces with much the same result, as long as inherently Coulombic (thus spherically symmetric, among other things), with the analysis as before. A variety of more complex potentials could be

109

considered but the essential quality is that there are asymptotic states and there are, maybe, bound states. We can largely determine the potential from ingoing asymptotic states that become "scattered" into outgoing asymptotic states (by the nonzero interaction potential), or in turn, verify our theoretical prediction for what that potential would be. This is where the "rubber meets the road" with theoretical physics connecting to experimental physics.

Note, when speaking of unbound asymptotic states, or free states, and bound states, we are speaking of two dynamical outcomes existing within the same dynamical system. We have seen this before, in the context of two-timing analysis and for perturbative analysis in general (perturbative analysis assumes the dynamics of a reference system, then considers a second system, the perturbed system). We can "see" the asymptotic states that are "free" from the interaction of interest, asymptotically, by capturing them in our detection apparatus. The same can't be said for the bound states, that we identify indirectly.

Let's recap the key questions, according to Reed and Simon [56], that scattering theory seek to answer (see [56] for further details). To get started, let's adopt their notation for free and bound states: ρ_+ is asymptotically free in the future ($t \to \infty$), ρ_- is asymptotically free in the past ($t \to -\infty$), and ρ is a bound state. From the Hamiltonian formulation we know we can speak of a "time transformation operator" acting on the aforementioned states with respect to a choice of Hamiltonian, here with/without interaction: $\{T_t, T_t^{(0)}\}$. Thus, it is possible to consider the asymptotic limits:

$$\lim_{t \to -\infty} \left(T_t \rho - T_t^{(0)} \rho_-\right) = 0 \quad and \quad \lim_{t \to \infty} \left(T_t \rho - T_t^{(0)} \rho_+\right) = 0.$$

(Eqn. 4-32)

These limits are only well-defined if solutions occur for pairs $\{\rho_-, \rho\}$ where for each ρ there is only one corresponding ρ_-, likewise for $\{\rho_+, \rho\}$. The key questions:

(1) What are the free states? Can they all be prepared experimentally (completeness on preparation)?
(2) Is there uniqueness on correspondence $\{\rho_-, \rho\}$ and $\{\rho_+, \rho\}$?
(3) Is there (weak) completeness on scattering? e.g., map all ρ_- onto $\rho \in \Sigma$, call that subset of Σ, Σ_{in}; repeat for ρ_+ to get Σ_{out}, does $\Sigma_{in} = \Sigma_{out}$? This is known as weak asymptotic completeness [56].

(4) Given the above, we can define a bijection of Σ onto itself, such that the following become well-defined: $\rho_- = \Omega^-\rho$ and $\rho_+ = \Omega^+\rho$, where Ω^- and Ω^+ are the bijective mappings. We can, thus, describe scattering in terms of a bijection:

$$S = (\Omega^-)^{-1}\Omega^+.$$

In classical mechanics this will always exist as a bijection on the phase space. In quantum mechanics S will be a linear unitary transformation known as the S-matrix.

(5) Are there symmetries? Sometimes S can be determined due to symmetries, this will be explored further in the QM context in [42].

(6) What is the analytic continuation? A common refinement for a Real theory, to encompass wave phenomena (such as in transitioning to a quantum theory), is to shift to a complex theory by seeing the Real theory as the boundary value of an analytic function. Analyticity of the S-transformation, according to choice, also imparts causality (as with the Feynman choice of contour integral definitions for propagators in [43]).

(7) Is it asymptotically complete: $\Sigma_{bound} + \Sigma_{in} = \Sigma_{bound} + \Sigma_{out}$? For classical mechanics the "+" operations are set theoretic so this reduces to the question of whether $\Sigma_{in} = \Sigma_{out}$ (weak asymptotic completeness) aside from a possible set of measure zero (i.e., there are set of measure zero issues – the set of bound states may be of measure zero with respect to the superset). In quantum theory the "+" is a direct sum of Hilbert spaces, which is more complicated and not discussed here.

Example 4.1. Classical Decay.

Consider a classical decay, A$\longrightarrow$ 3B, in which the first particle decays to three identical particles of mass m. Suppose that each final particle has the same energy in the CM frame, that the original particle moves with speed V along the lab's z-axis, and that the decay energy is ϵ. If one of the particles emerges along the positive z-axis, at what angle to the z-axis do the other two particles emerge?

Solution

We have the same energy in CM frame, i.e., same momentum. Thus, in CM frame

$$\frac{1}{2}(3m)V^2 = 3\frac{1}{2}(m)V'^2 + \epsilon \rightarrow (mV') = \sqrt{m^2V^2 - \frac{2}{3}m\epsilon}$$

and

$$\tan\phi = \frac{|(m\vec{V}')|\sin(60°)}{|(3m\vec{V})| - |(m\vec{V}')|\cos(60°)} \quad where \ \sin 60° = \frac{\sqrt{3}}{2} \quad \cos 60°$$

$$= \frac{1}{2}$$

Thus,

$$\phi = \tan^{-1}\left\{\frac{\sqrt{m^2V^2 - \frac{2}{3}m\epsilon}\,\frac{\sqrt{3}}{2}}{3mV - \sqrt{m^2V^2 - \frac{2}{3}m\epsilon}\,\frac{1}{2}}\right\}$$

$$= \tan^{-1}\left\{\frac{\sqrt{3m^2V^2 - 2m\epsilon}}{6mV - \sqrt{m^2V^2 - \frac{2}{3}m\epsilon}}\right\}$$

Exercise 4.1. Classical Decay.

Example 4.2. (F&W 1.14)
Consider Rutherford scattering off of a nuclear surface when the cross-section to strike the nuclear surface is $\sigma_r = \pi b^2$ for impact parameter at minimum r: $r_{min} = b$. Recall that the system energy asymptotically, with ingoing velocity V_∞, is simply

$$E = \frac{1}{2}mV_\infty^2 \quad \rightarrow \quad V_\infty = \sqrt{\frac{2E}{m}}.$$

We also have for (conserved) angular momentum:
$$M_\theta = mV_\infty b = \sqrt{m2E}b.$$
Thus, the effective potential with indicated M_θ and Coulomb potential $V_c = \frac{zZe^2}{R}$ is:

$$U_{eff} = \frac{M_\theta^2}{2mR^2} + V_c = E \ \rightarrow \ \frac{m2Eb^2}{2mR^2} + V_c = E \ \rightarrow \ b^2 = R^2\frac{(E - V_c)}{E}$$

Thus,

$$\sigma_r = \pi b^2 = \pi R^2(1 - V_c/E).$$

Related Exercises: see Fetter&Walecka [29].

Example 4.3. (F&W 1.17)
Consider scattering off of the potential

$$V(r) = \begin{cases} 0 & r > a \\ -V_0 & r < a \end{cases}$$

(1) Show orbit is identical to a light ray refracted by a sphere of radius a and $= \sqrt{(E + V_0)/E}$.

(2) Find the differential elastic cross section.

Solution

(1) Recall $F 2\pi b\, db = F d\sigma_a(\theta)$ and $d\Omega = 2\pi \sin\theta\, d\theta \Rightarrow \dfrac{d\sigma}{d\Omega} = \dfrac{b}{\sin\theta}\left|\left(\dfrac{db}{d\theta}\right)\right|$

Have: $mV_1 \sin\theta_1 = mV_2 \sin\theta_2$ and $E = \dfrac{P_1^2}{2m} + U_1 = \dfrac{P_2^2}{2m} + U_2$. Thus:

$$\sin\theta_1 = \sin\theta_2 \sqrt{1 + \frac{2}{mV_1^2}V_0} \;\rightarrow\; \sin\theta_1 = \sqrt{(E + V_0)/E}\,\sin\theta_2$$

Thus, the orbit is identical to a light ray refracted by a sphere of radius a and $n = \sqrt{(E + V_0)/E}$

$$\sin\theta_2 = \frac{\sin\theta_1}{\sqrt{(E + V_0)/E}}$$

Deflection angle corresponding to θ_1 and θ_2 is $\theta = (\theta_1 - \theta_2)$. Thus, $\theta_1 = \dfrac{\theta}{2} + \theta_2$, and since $b = a \sin\theta_1$ we have:

$$\sin\theta_1 = \sin\left\{\frac{\theta}{2} + \theta_2\right\} = \sin\left(\frac{\theta}{2}\right)\sin\theta_2 + \cos\left(\frac{\theta}{2}\right)\cos\theta_2 = \frac{\sin\left(\frac{\theta}{2}\right)\sin\theta_1}{n} +$$
$$\cos\left(\frac{\theta}{2}\right)\sqrt{1 - \sin^2\theta_1^2}$$

$$\sin^2\theta_1 = \frac{\sin^2\left(\frac{\theta}{2}\right)}{\left(\frac{1}{n} - \cos\left(\frac{\theta}{2}\right)\right)^2 + \sin^2\left(\frac{\theta}{2}\right)}$$

$$b^2 = a^2 \sin^2\theta_1 = \frac{a^2 n^2 \sin^2\left(\frac{\theta}{2}\right)}{+ n^2 \sin^2\left(\frac{\theta}{2}\right) + \left(1 - 2n\cos\left(\frac{\theta}{2}\right) + n^2 \cos^2\left(\frac{\theta}{2}\right)\right)} = \frac{a^2 n^2 \sin^2\left(\frac{\theta}{2}\right)}{1 + n^2 - 2n\cos\left(\frac{\theta}{2}\right)}$$

$$2bdb = a^2n^2 \left\{ \frac{2\sin\left(\frac{\theta}{2}\right)\cdot\frac{1}{2}\cos\left(\frac{\theta}{2}\right)}{1+n^2-2n\cos\left(\frac{\theta}{2}\right)} \right.$$

$$\left. +\frac{(-1)a^2n^2\sin^2\left(\frac{\theta}{2}\right)\left[-2n\left(-\frac{1}{2}\sin\frac{\theta}{2}\right)\right]}{(\quad)^2} \right\}$$

$$= \frac{a^2n^2}{\left(1+n^2-2n\cos\left(\frac{\theta}{2}\right)\right)^2}\left\{ \sin\left(\frac{\theta}{2}\right)\cos\left(\frac{\theta}{2}\right)\left(1+n^2-2n\cos\frac{\theta}{2}\right) - \right.$$

$$\left. n\sin^3\left(\frac{\theta}{2}\right) \right\}$$

Thus,

$$\frac{d\sigma}{d\Omega} = \frac{b}{\sin\theta}\left|\frac{db}{d\theta}\right|$$

$$= \frac{a^2n^2}{4\cos\left(\frac{\theta}{2}\right)\left(1+n^2-2n\cos\left(\frac{\theta}{2}\right)\right)^2}\left\{\cos\left(\frac{\theta}{2}\right)(1+n^2)\right.$$

$$\left. -2n+n\left(1-\cos^2\left(\frac{\theta}{2}\right)\right)\right\}$$

$$\frac{d\sigma}{d\Omega} = \frac{a^2n^2}{4\cos\left(\frac{\theta}{2}\right)\left(1+n^2-2n\cos\left(\frac{\theta}{2}\right)\right)^2}\left\{\left(n\cos\left(\frac{\theta}{2}\right)-1\right)\left(n\right.\right.$$

$$\left.\left. -\cos\left(\frac{\theta}{2}\right)\right)\right\}$$

Related Exercises: see Fetter&Walecka [29].

Example 4.4. (F&W 1.18)
Consider a small particle at large impact parameter b from central
potential V(r) with only a slight deflection occurring upon scattering.
(a) Use an impulse approximation to derive the small deflection angle.
(b) Examine the case $V(r) = \gamma r^{-n}$, where both γ and n are positive.
(c) Examine the case $V(r) = \gamma e^{-\lambda r}$.
(d) In QM the small angle part of the cross-section is different than
classical, discuss.

Solution

(a) In the impulse approximation we have $\theta_1 \approx \dfrac{P'_{1y}}{m_1 v_\infty}$ and $P'_{1y} =$

$\int_{-\infty}^{\infty} F_y\, dt = \int_{-\infty}^{\infty} -\dfrac{dU}{dr}\dfrac{y}{r}\, dt$

Assume small deflection $y = b$, $dt = \dfrac{dx}{v_\infty}$:

$$\theta = \frac{b}{m_1 v_\infty^2} \int_{-\infty}^{\infty} -\frac{dU}{dr}\frac{dx}{r} = \frac{2b}{m_1 v_\infty^2}\left|\int_b^{\infty} \frac{dU}{dr}\frac{dr}{\sqrt{r^2-b^2}}\right|$$

(b) $V(r) = \gamma r^{-n}$ $r > 0, n > 0$

$$\theta = \frac{2b}{m_1 v_\infty^2}\left|\int_b^{\infty} \gamma(-n)r^{-n-1}\frac{dr}{\sqrt{r^2-b^2}}\right| = \frac{2b}{m_1 v_\infty^2}n\gamma\left|\int_b^{\infty} \frac{r^{-(n-1)}dr}{\sqrt{r^2-b^2}}\right|$$

$$\theta = \frac{2b}{m v_\infty^2} \int_b^{\infty} \frac{dr}{\sqrt{r^2-b^2}}\gamma n r^{-n-1} = \frac{2b}{m v_\infty^2} \int_1^{\infty} \frac{\gamma n b\, dx\, b^{-(n+1)} x^{-(n+1)}}{b\sqrt{x^2-1}}$$

$$= \frac{2b}{m v_\infty^2 b^n} \int_1^{\infty} \frac{x^{-(n+1)}}{\sqrt{x^2-1}}dx$$

Thus, $\theta = \dfrac{C}{b^n}$ $C = \dfrac{2}{m v_\infty^2}\int_1^{\infty} \dfrac{x^{-(n+1)}}{\sqrt{x^2-1}}dx$.

So,

$$\frac{d\theta}{db} = \frac{-nC}{b^{n+1}} \quad and \quad \frac{d\sigma}{d\Omega} = \frac{1}{nC}\frac{b^{n+2}}{\sin\theta} \cong \frac{1}{nC}\frac{b^{n+2}}{\theta}$$

Thus,

$$b^{n+2} = \left(\frac{C}{\theta}\right)^{\left(\frac{n+2}{n}\right)} \quad and \quad \frac{d\sigma}{d\Omega} = C'\theta^{-\left(2+\frac{2}{n}\right)}.$$

For $n = 1$, $\dfrac{d\sigma}{d\Omega} \simeq C'\theta^{-4}$ ← Rutherford: $\left(\dfrac{d\sigma}{d\Omega}\right)_{el} = \left(\dfrac{zZe^3}{4E\sin^2\frac{1}{2}\theta}\right)^2$

$n = 2$, $\dfrac{d\sigma}{d\Omega} \simeq C'\theta^{-3}$ ← $\left(\dfrac{d\sigma}{d\Omega}\right)_{el} = \dfrac{\gamma\pi^2}{E\sin\theta}\dfrac{\pi-\theta}{\theta^2(2\pi-\theta)^2}$

115

For σ_τ to be well defined: $\int \frac{d\sigma}{d\Omega} d\Omega < \infty$. Here we have:

$$\int_0^\theta C' \theta^{-\left(2+\frac{2}{n}\right)} d\Omega \sim \int_0^\theta C' \theta^{-\left(2+\frac{2}{n}\right)} \theta d\theta \sim \theta^{-\frac{2}{n}}\Big|_0^\theta = \infty \text{ for } n > 0$$

So, the cross-section is only well-defined if n<0.

(c) Have: $V(r) = \gamma e^{-\lambda r}$ $\quad$ $r = bx$

$$\theta = \frac{2b}{m_1 v_\infty^2} \left| \int_b^\infty -\frac{\gamma \lambda e^{-\lambda r} dr}{\sqrt{r^2 - b^2}} \right| = b^2 \left(\frac{\lambda 2\lambda}{m_1 v_\infty^2}\right) \int_1^\infty \frac{x e^{-\lambda bx} dx}{\sqrt{x^2 - 1}}$$

Consider $b\lambda \gg 1$ only $x \approx 1$ contributes

$$\theta = \gamma b\lambda \left(\frac{2}{m_1 v_\infty^2}\right) \int_1^\infty \frac{e^{-\lambda b}\, e^{-\lambda b\epsilon}}{\sqrt{2}\sqrt{\epsilon}} d\epsilon = \gamma b e^{-\lambda b} K \qquad K$$

$$= \left(\frac{\sqrt{2}\lambda}{m_1 v_\infty^2}\right) \int_1^\infty \frac{e^{-\lambda b\epsilon}}{\sqrt{\epsilon}} d\epsilon$$

Thus,

$$\theta = \gamma \sqrt{\frac{\pi b}{\lambda}} e^{-\lambda b} \left(\frac{\lambda}{m_1 v_\infty^2}\right).$$

Since

$$\log \theta \approx -\lambda b \quad \rightarrow \quad b \sim \lambda^{-1} \log\left(\frac{1}{\theta}\right) \quad \rightarrow \quad \frac{d\sigma}{d\Omega} \sim \frac{b}{\theta}\frac{db}{d\theta}$$

Thus, σ_τ not well defined because $\int_0^x \frac{dx}{x \log x} = \log(\log x)_{x\to\infty} \to \infty$

(d) Classically: no zero-angle scattering for finite b; while, QM has finite probability density for zero-angle scattering.

Related Exercises: see Fetter&Walecka [29].

Chapter 5. Collective Motion

Brief mention will now be given to collective motion for idealized cases such as rigid bodies and simple material bodies, with the phenomenological discussion involving material bodies partly left to Ch. 8 Phenomenology & Dimensional Analysis. This brief review starts with Rigid Body motion.

5.1 Rigid Body Motion

For a rigid body all of the internal loadings are net zero. If the geometry of a rigid body is static then forces applied must be balanced, and transmitted, through the rigid body such that the net forces and torsions are net zero. At any position in the body we can evaluate the net forces and moments of force according to six scalar equations of equilibrium:

$$\Sigma F_x = 0, \Sigma F_y = 0, \Sigma F_z = 0, \Sigma M_x = 0, \Sigma M_y = 0, \Sigma M_z = 0.$$
(Eqn. 5-1)

When speaking of a homogeneous material comprising the rigid body, it is possible to speak of the average normal stress to a cross-sectional surface ($\sigma = N/A$, where N is the internal axial load and A is the cross-sectional area) and the average shear stress to a cross-sectional surface ($\tau_{avg} = S/A$, where S is the shear force acting on the cross-section A). Let's consider some classic problems from Hibbeler [59,60] to work through some of these Statics issues and see their application.

Example 5.1. (Hibbeler 1-12)

A beam is held horizontally with its left end at a wall-mounted pin (point A). Proceeding left to right along the beam we have points labeled as follows: 1ft to the right of A there is point D, another 2 ft and point B, another 1 ft and point E, another 2 ft and point G, and another 1ft to reach the end where a load is indicated due to a cable connection at 30 degrees outward (rightward) from the vertical. At point B is a support beam, directed upwards to the wall, forming a 3-4-5 triangle with the wall (upper pin mount labeled C), where the 3 corresponds with the 3ft from A to B. The load on the cable is 150 lb. There is also a uniform distributed load between point B and the end of the beam of 75 lb/ft. Along the diagonal support beam, down 1ft from the support pin at point C, is an internal beam point labeled F.

"Determine the resultant internal loadings at cross sections at points F and G on the assembly."

Consider the free diagram for the horizontal beam, this will allow us to solve for the axial beam force F_{CB} from which the internal loading at F can be trivially obtained. A cut (sectioning) to a free-body at the cross-section of G is taken to the right side for another simple free body analysis to get the internal loading at G. First, for F_{CB}:

$$\sum M_A = 0 \rightarrow 3(0.8)F_{BC} - 5(300) - 7(150)(0.5)\sqrt{3} = 0 \rightarrow F_{BC}$$
$$= 1{,}003.9 \ lb.$$

From this we have to the internal load at F:

$$N_F = F_{BC} = 1{,}003.9 \ lb, \quad S_F = 0, \quad and \quad M_F = 0.$$

Let's now consider the internal loading at G by way of the free-body section (see [59,60] for details) consisting of the body on the right side of the cut:

$$\sum M_G = 0 \rightarrow M_G - (0.5)(75) - (1)(150)(0.5)\sqrt{3} = 0 \rightarrow M_G$$
$$= 167.4 ft \ lb .$$

$$\sum F_x = 0 \rightarrow N_G + 150(0.5) = 0 \rightarrow N_G = -75 lb.$$

$$\sum F_y = 0 \rightarrow V_G - 75 - 150(0.5)\sqrt{3} = 0 \rightarrow N_G = 205 lb$$

Exercise 5.1. *Redo with 150 lb → 250 lb.*

Example 5.2. Hibbeler (1-66)

A "frame" is formed by a vertical wall and two beams coming together to form a 3-4-5 triangle (hypotenuse upwards, so beam under tension, not compression). The wall mounts are hinged pins, as is the connection between the beams. The distance between the wall mounts (vertical length) is 2m and the horizontal beam has length 1.5m. The lower wall-mount is labeled point A, the upper B, and the connection point of the beams is point C. Thus, the Hypotenuse is length BC. At point C a load P is indicated vertically downward. Cutting through beam BC vertically is indicated a cross-sectional cut labeled "a-a".

"Determine the largest load **P** that can be applied to the frame without causing either the average normal stress or the average shear stress at section a-a to exceed $\sigma = 150MPa$ and $\tau = 60MPa$, respectively. Member CB has a square cross section of 25mm on each side.

118

Let's start with consideration of the horizontal beam as a free body to obtain F_{BC} in terms of **P**:

$$\sum M_A = 0 \quad \rightarrow \quad 0.8F_{BC} = P.$$

(Eqn. 5-2)

The cross-section under consideration is not orthogonal to the axis of the beam, thus need to correct the normal force and (non-zero) shear force accordingly:

$$N_{aa} = 0.6F_{BC} = 0.75P \quad and \quad S_{aa} = 0.8F_{BC} = P.$$

The area of the cross-section is: $A_{aa} = A/\cos\theta = (5/3)A$. Thus the normal stress to the indicated a-a cross-section is maximal when at the stress limit indicated:

$$\sigma = \frac{N_{aa}}{A_{aa}} = 150MPa \rightarrow P_{max} = 208kN.$$

(Eqn. 5-3)

The maximum load P that can be according to the normal stress is limited to be $P_{max} = 208kN$.

The shear stress indicated at a-a can be at most 60MPa from which we calculate:

$$\tau = \frac{S_{aa}}{A_{aa}} = 60MPa \rightarrow P_{max} = 22.5kN.$$

(Eqn. 5-4)

The maximum load P that can be according to the shear stress is limited to be $P_{max} = 22.5kN$, and since this limit is reached sooner, the maximum load possible at P is 22.5kN (to avoid shear failure).

Let's consider some dynamical situations with rigid bodies (a few have already been mentioned, but with idealized massless rods).

Exercise 5.2. Redo with $\sigma = 250MPa$.

Example 5.3. A plank leaning against a wall.
Let's consider the problem of a plank leaning against a wall. If the plank makes an angle θ_0 with the floor, initially, and the plank is free to slide along the floor (no friction), what is its motion? When, if ever, does the plank leave contact with the wall? When, if ever, does the plank leave contact with the floor? This is similar to problem 3.18 on pg 85 of [29], with plank of length L and mass M.

To begin recall that the moment of inertia of a (uniform) plank about its center of mass is $I = \frac{1}{12}ML^2$. The kinetic energy term can then be given in terms of the center-of-mass linear motion and the rotation about that center:

$$T = \frac{1}{2}M(\dot{x}^2 + \dot{y}^2) + \frac{1}{2}I\dot{\theta}^2,$$

where the (x, y) coordinates of the center-of-mass are related to θ by $x = \frac{L}{2}\cos\theta$ and $y = \frac{L}{2}\sin\theta$ (while maintaining contact with the wall). The potential energy is simply: $V = Mgy$. The Lagrangian is, thus:

$$L = \frac{1}{2}M(\dot{x}^2 + \dot{y}^2) + \frac{1}{2}I\dot{\theta}^2 - Mgy \rightarrow L$$

$$= \frac{1}{2}M\left(\frac{L}{2}\right)^2 \dot{\theta}^2 + \frac{1}{2}I\dot{\theta}^2 - Mg\frac{L}{2}\sin\theta$$

The Euler-Lagrange (E-L) equation for the latter (constrained form) then gives:

$$\dot{\theta}^2 = \frac{3g}{l}(\sin\theta_0 - \sin\theta).$$

Since we are interested in the contact constraints (and when they fail), let's revert to the initial form, and add Lagrange multipliers for the constraints:

$$L(\lambda, \tau) = \frac{1}{2}M(\dot{x}^2 + \dot{y}^2) + \frac{1}{2}I\dot{\theta}^2 - Mgy + \tau\left(x - \frac{L}{2}\cos\theta\right)$$
$$+ \lambda\left(y - \frac{L}{2}\sin\theta\right).$$

The equations of motion for the (x, y) coordinates of the center-of-mass and the (λ, τ) Lagrange multipliers for the x-constraint are:

$$M\ddot{x} - \tau = 0 \quad \rightarrow \quad \tau = -\frac{ML}{2}\left(\cos\theta\,\dot{\theta}^2 + \sin\theta\,\ddot{\theta}\right)$$
$$= \frac{3gM}{2}\cos\theta\left(\frac{3}{2}\sin\theta - \sin\theta_0\right)$$

where the τ multiplier goes to zero when:

$$\frac{3}{2}\sin\theta_c - \sin\theta_0 = 0.$$

Thus, the plank leaves the wall when the contact point is at height:

120

$$Y = 2y = 2\left(\frac{L}{2}\right)\sin\theta_C = \frac{2}{3}L\sin\theta_0.$$

At the instant the ladder leaves the wall the x coordinate is free and has:

$$x = \frac{L}{2}\sqrt{1 - \left(\frac{2}{3}\right)^2 \sin^2\theta_0} \quad and \quad \dot{x} = -\frac{\sqrt{gL}}{3}(\sin\theta_0)^{\frac{3}{2}} \quad and \quad \ddot{x} = 0$$

Let's now examine the y-constraint before and after the plank leaves the wall:

$$M\ddot{y} + Mg - \lambda = 0 \quad \rightarrow \quad \lambda = \frac{ML}{2}\left(-\sin\theta\,\dot{\theta}^2 + \cos\theta\,\ddot{\theta}\right) + Mg$$

Before the plank leaves the wall we have $\dot{\theta}^2 = \frac{3g}{L}(\sin\theta_0 - \sin\theta)$ and $\ddot{\theta} = -\frac{3g}{2L}\cos\theta$, for which $\lambda > 0$ always. After the plank leaves the wall we have $\dot{\theta}^2 = \frac{g}{L}\sin\theta_0$ and $\ddot{\theta} = 0$, for which $\lambda > 0$ always. Thus λ never goes to zero, and the plank never leaves the floor, with c.o.m. y-motion similarly expressed as with the x-motion above.

Exercise 5.3. Suppose there is a worker on the ladder at midpoint, of mass M, repeat the analysis.

Example 5.4. Revolving pipe, at fixed angle, with ball inside.
Consider a pipe that revolves with constant angular velocity ω about a vertical axis forming a fixed angle α with it. Inside the pipe is a ball of mass m that slides freely without friction. Using spherical coordinates, at time t=0 let the ball position be $r = a$ and $\frac{dr}{dt} = 0$. For all times of interest the ball remains in the upper part of the pipe. (a) Find the Lagrangian; (b) Find the equations of motion; (c) Find the constants of the motion; (d) Find t as a function of r in the form of an integral.

Solution
(a) The Lagrangian for the ball's motion is given by
$$L = \frac{1}{2}m\left(\frac{ds}{dt}\right)^2 - mgr\cos\alpha$$

where, for spherical coordinates: $ds^2 = dr^2 + r^2(d\theta^2 + sin^2\theta d\varphi^2)$. Thus,

$$L = \frac{1}{2}m\left(\dot{r}^2 + r^2(\dot{\theta}^2 + sin^2\theta\dot{\varphi}^2)\right) - mgrcos\alpha, \quad with \quad \theta = \alpha, \quad \dot{\varphi} = \omega$$

and we get:

$$L = \frac{1}{2}m(\dot{r}^2 + r^2sin^2\alpha\omega^2) - mgrcos\alpha$$

(b) The equation of motion for r for fixed rotation frequency and specified angle of declination:

$$m\ddot{r} - mrsin^2\alpha\omega^2 + mgcos\alpha = 0 \rightarrow \frac{d}{dt}\left\{\frac{1}{2}\dot{r}^2 - \frac{1}{2}r^2sin^2\alpha\omega^2 + rgcos\alpha\right\}$$
$$= 0.$$

(c) The constant of the motion is thus

$$\dot{r}^2 - r^2sin^2\alpha\omega^2 + r2gcos\alpha = const$$

From r=a and $\frac{dr}{dt} = 0$ initialization we have

$$const = 2agcos\alpha - (a\omega sina)^2.$$

(d) We can write

$$\left(\frac{dr}{dt}\right)^2 = \dot{r}^2 = 2gcos\alpha(a - r) + (\omega sin\alpha)^2(r^2 - a^2)$$

or, switching to integral form:

$$dt = \frac{dr}{\sqrt{2gcos\alpha(a - r) + (\omega sin\alpha)^2(r^2 - a^2)}}$$

Thus,

$$t = \int \frac{dr}{\sqrt{2gcos\alpha(a - r) + (\omega sin\alpha)^2(r^2 - a^2)}}.$$

Exercise 5.4. *Repeat analysis for a revolving paraboloid curved pipe with ball inside.*

5.2 Material Bodies

Thus far we've seen how to compute stress as a Force over an area ($\sigma = F/A$). With non-idealized bodies (such as rigid bodies), i.e., material bodies, there will be a response, a deformation, to this stress. To quantify this deformation let's define strain:

$$\epsilon = \frac{\Delta L}{L}.$$

(Eqn. 5-5)

122

The relation between applied normal stress and resulting strain deformation is given by Hooke's Law:

$$\sigma = Y\epsilon,$$

(Eqn. 5-6)

where Y is a constant appropriate to the material under consideration known as Young's modulus. From this we can compute the strain energy density: $u = \sigma\epsilon/2$. Similar relations exist for shear stress. If we consider a constant load and cross-sectional area we can group the equations to get a relation on the change in length for given applied (normal) force:

$$\delta = \frac{FL}{AY}.$$

(Eqn. 5-7)

If there are connected sections with different areal cross-sections, etc., their δ's are additive.

Lastly, for this brief overview of material bodies, is to account for thermal stress (most thermal effects are not discussed until [44]). It is well known that material bodies expand or contract under change of temperature. This is described by the following:

$$\delta_T = \alpha\Delta TL,$$

(Eqn. 5-8)

where α is the linear coefficient of thermal exapansion.

Example 5.5. Hibbeler (3-8)
A beam is held horizontally, initially, with length $10ft$, and a distributed load on its entirety of w. It is held at on end by a (wall-mounted) hinged pin and at the other end by a guy wire support at 30 degrees to the horizontal.

"The rigid beam is supported by a pin at C and an A-36 guy wire AB. If the wire has a diameter of 0.2in., determine the distributed load w if the end B is displaced 0.75in. downward."

We need to first compute the strain on the guy wire and from this determine what load is present. The original length AB is 11.547ft. The stretched length of the guy wire is 11.578ft, thus the strain is $\epsilon = 0.00269$. The Young's modulus for A-36 guy wire is $29x10^3 ksi$, thus have:

$$\frac{F}{A} = Y\epsilon \;\;\rightarrow\;\; F = 2.45kip \;\;\rightarrow\;\; w = \frac{0.245kip}{ft}.$$

123

Exercise 5.5. Redo for wire diameter 0.3in, and displacement of end B is 1.0in along length AB.

Example 5.6. Hibbeler (4-70)
A rod is horizontally mounted between two walls by use of two (identical) springs on either end, between the wall and the rod's ends.

"The rod is made of A992 steel [$\alpha = 6.6x10^{-6}/°F$] and has a diameter of 0.25 in. If the rod is 4ft long when the springs [$k = 1000lb/in$] are compressed 0.5 in. and the temperature of the rod is $T = 40°F$, determine the force in the rod when its temperature is $T = 160°F$."

From $\delta_T = \alpha \Delta T L \rightarrow \delta_T = 3.168 \times 10^{-3} ft$. With the two springs acting together we have the force acting inward on both sides of:
$$F = k\left(\frac{\delta_T}{2}\right) = 19 \; lb.$$

Exercise 5.6. Repeat for T $= 360°F$ and spring compression 0.75in.

5.3 Hydrostatics and Stationary Fluid Flow
Hints of Special Relativity: Fizeau, the Relativist Doppler Effect, and the Bondi K-calculus
Special Relativity is revealed when going to field theory to describe EM. Hints of the existence of SR for consistencies-sake are seen in primitive early experiments with light, but significance not understood at the time.

Fizeau 1851 [22] found that the speed of light in water moving with a speed v (relative to the lab) could be expressed as:
$$u = \frac{c}{n} + kv,$$
(Eqn. 5-9)
where the "dragging coefficient" was measured to be $k = 0.44$. The value of k predicted by Lorentz velocity addiction:
$$x = \frac{x' + vt'}{\sqrt{1 - \frac{v^2}{c^2}}} \rightarrow u_x = \frac{dx' + vdt'}{dt' + \frac{v}{c^2}dx'} = \frac{u_x' + v}{1 + \frac{v}{c^2}u_x'}$$

124

Treating light as a particle, the laboratory observer will find its speed to be:

$$u_x = \frac{c/n + v}{1 + \frac{v}{c^2}\frac{c}{n}} \cong \frac{c}{n} + \left(1 - \frac{1}{n^2}\right)v.$$

Water has $n \cong 4/3$, thus:

$$u_x \cong \frac{c}{n} + (0.44)v,$$

thus agreement with the experiment done in 1851.

Chapter 6. Legendre Transformation and the Hamiltonian

Let's start with the Lagrangian and perform a Legendre transformation to get the Hamiltonian formulation:

$$dL = \sum_i \frac{\partial L}{\partial q_i} dq_i + \frac{\partial L}{\partial \dot{q}_i} d\dot{q}_i$$

Substituting the relation for generalized momenta, $p_i = \frac{\partial L}{\partial \dot{q}_i}$, and

Lagrange's equations: $F_i = \dot{p}_i = \frac{\partial L}{\partial q_i}$,

$$dL = \sum_i \dot{p}_i dq_i + p_i d\dot{q}_i.$$

Regrouping we arrive at the Hamiltonian of the system (seen earlier as the energy if the system is conserved):

$$dH = d\left(\sum_i p_i \dot{q}_i - L\right) = -\sum_i \dot{p}_i dq_i + \dot{q}_i dp_i,$$

(Eqn. 6-1)

which indicates that, $\dot{p}_i = -\frac{\partial H}{\partial q_i}$, and $\dot{q}_i = \frac{\partial H}{\partial p_i}$.

Now consider the total time-derivative of the Hamiltonian:

$$\frac{dH}{dt} = \frac{\partial H}{\partial t} + \sum_i \frac{\partial H}{\partial q_i} \dot{q}_i + \frac{\partial H}{\partial p_i} \dot{p}_i = \frac{\partial H}{\partial t}$$

(Eqn. 6-2)

and if H not explicitly time-dependent we get $\frac{dH}{dt} = 0$, thus $H = E$, for constant E, the conserved energy of the system.

6.1 Area Conserving Mappings
Let's consider infinitesimal motion of an object in terms of the generalized coordinates going from (q_0, p_0) to (q_1, p_1) in phase space:

$$q_1 = q_0 + \delta t \dot{q}|_{q=q_0} + O(\delta t^2) = q_0 + \delta t \frac{\partial H(q_0, p_0, t)}{\partial p_0} + O(\delta t^2)$$

$$p_1 = p_0 + \delta t \dot{p}|_{p=p_0} + O(\delta t^2) = p_0 - \delta t \frac{\partial H(q_0, p_0, t)}{\partial q_0} + O(\delta t^2)$$

Viewed as a coordinate transformation, the Jacobian is:

$$\frac{\partial(q_1, p_1)}{\partial(q_0, p_0)} = \begin{vmatrix} \dfrac{\partial q_1}{\partial q_0} & \dfrac{\partial p_1}{\partial q_0} \\[2mm] \dfrac{\partial q_1}{\partial p_0} & \dfrac{\partial p_1}{\partial p_0} \end{vmatrix} = 1 + O(\delta t^2).$$

(Eqn. 6-3)

As the infinitesimal is taken to zero we see that any flow satisfying Hamilton's equations is area-preserving (Jacobian=1). The converse is true as well, if the flow an enclosed region under the phase space mapping or flow is area preserving, then the flow satisfies Hamilton's equations.

6.2 Hamiltonians & Phase Maps

Since the Hamiltonian is conserved, it involves motion in phase space along curves of constant $H = E$. The phase diagram for a Hamiltonian system, thus, consists of contours of constant H, like a contour map. Previously,

$$L = \frac{1}{2} m \dot{q}^2 - U(q) \rightarrow E = \frac{1}{2} m \dot{q}^2 + U(q)$$

(Eqn. 6-4)

using,

$$H = \sum_i p_i \dot{q}_i - L, \text{ with } p_i = \frac{\partial L}{\partial \dot{q}_i}$$

(Eqn. 6-5)

Now have:

$$H(p, q) = \frac{p^2}{2m} + U(q).$$

(Eqn. 6-6)

The contours, or level curves, of the Hamiltonian, are invariant sets, as are fixed points. Fixed points in the phase space occur when the gradient of the Hamiltonian is zero: $\nabla H = 0$, $i.e.$ $\partial H/\partial q = 0$, and $\partial H/\partial p = 0$. The system is at equilibrium when at a fixed point so identifying these points, and related attractors and limit cycles, will thus be of interest in understanding a systems dynamics and asymptotic behavior (all to be discussed).

Cases I-VI in what follows describe instances of ordinary differential equations (ODE's), with stability as indicated. A complete analysis along

128

these lines, locally, reveals the various types of stability, and general criteria [31] and is discussed in the section after this. If a fully global separability can be obtained, it is clearest in the Hamiltonian-Jacobi formalism (also discussed in a later section).

Let's start with an analysis of second order autonomous systems along the lines of [28]. This covers many systems of interest, as well as the linearized (local) approximation for any system. We begin by describing the system via a real vector, $r(t)$ with 2N components if there are N degrees of freedom, with an associated "phase velocity" $\dot{r}(t) = v(t)$, which is aa first-order vector differential equation. The order is defined as the minimum number of coupled first-order equations, here 2N.

The motions of a second order system can be described in terms of the flow lines, and fixed points (if any), in their associated $\{r(t), v(t)\}$ "phase portrait" or "phase diagram". This allows a qualitative analysis of the properties of a system, where the special cases analyzed in cases I-VI provide an understanding of the building blocks in such a qualitative analysis.

Following [28], let's first consider phase space maps for special, lowest-order q, $U(q)$ cases, then describe a general class of potentials arrived at by construction from those special cases. To begin, consider $U(q) = aq$:

Example 6.1. Case I. $U(q) = aq$. The Uniform Force Field. $aq = E - \frac{p^2}{2m}$:

Recall that $\dot{p}_i = -\frac{\partial H}{\partial q_i}$, and $\dot{q}_i = \frac{\partial H}{\partial p_i}$ and suppose $p = 0$ at t_0 and q_0:

$$H(p, q) = \frac{p^2}{2m} + aq \rightarrow \dot{p} = -a \quad and \quad \dot{q} = \frac{p}{m}$$

Integrating the first-order equations:

$$p = -a(t - t_0) \ and \ q = q_0 - \frac{a}{2m}(t - t_0)^2.$$

Exercise 6.1. Show the phase space map for Hamiltonian with potential $U(q) = aq$ (and graph of potential). Show that there are no fixed points.

Example 6.2. Case II. $U(q) = +\frac{1}{2}aq^2$. The Linear Oscillator. $\frac{1}{2}aq^2 + \frac{p^2}{2m} = E$ (circles/ellipses in phase space):

$$H(p,q) = \frac{p^2}{2m} + \frac{1}{2}aq^2 \rightarrow \dot{p} = -aq \ \text{and} \ \dot{q} = \frac{p}{m}$$

The second-order equation of motion that results is:

$$\ddot{q} = -\frac{a}{m}q = -\omega^2 q \rightarrow q = A\cos(\omega t + \delta) \rightarrow p = -m\omega A \sin(\omega t + \delta).$$

This is classic simple harmonic motion with period $T = 2\pi/\omega$ and $E = \frac{1}{2}mA^2\omega^2$.

Exercise 6.2. Show the phase space map for Hamiltonian with potential $U(q) = +\frac{1}{2}aq^2$ (along with graph of potential). Show that the level curves are ellipses and that there is an elliptic fixed point at q=0, p=0.

Example 6.3. Case III. $U(q) = -\frac{1}{2}aq^2$. The Linear Repulsive Force (Quadratic Potential Barrier).

$$H(p,q) = \frac{p^2}{2m} - \frac{1}{2}aq^2 \rightarrow \dot{p} = aq \ \text{and} \ \dot{q} = \frac{p}{m}$$

The second-order equation of motion that results is:

$$\ddot{q} = \frac{a}{m}q = \gamma^2 q \rightarrow q = Ae^{\gamma t} + Be^{-\gamma t} \rightarrow p$$
$$= m\gamma Ae^{\gamma t} - m\gamma Be^{-\gamma t}, \text{and } E = -2m\gamma^2 AB.$$

So far we've seen a case with no fixed point, an elliptic fixed point, and a hyperbolic fixed point. These are some of the main categories of interest, but to be complete, let's consider a system described by a vector function of time $r(t) = (q(t), p(t))$ that satisfies a first-order vector differential equation of motion:

$$\frac{dr(t)}{dt} = (\dot{q}(t), \dot{p}(t)) = v(q, p, t)$$

A point (q, p) where $v(q, p, t) = 0$ is known as a fixed point, it represents the system in equilibrium. If as $t \rightarrow \infty$ we have $r(t) \rightarrow r_0$, then r_0 is called an attractor. A strong attractor occurs when a phase trajectory anywhere in some neighborhood of the attractor point r_0 results in the trajectory joining (asymptoting to) the attractor.

Separation of variables is generally possible, from the theory of ordinary differential equations [32], and stability [31], and will be used to categorize the types of flows (with or without stable points) in the remainder of this section (along the lines of [28]). Further discussion of separability occurs in a later section where the Hamilton-Jacobi equation is discussed [27].

Exercise 6.3. Show the phase space map for Hamiltonian with potential $U(q) = -\frac{1}{2}aq^2$. Show the level curves are hyperbolas, or straight-lines if degenerate case (show the separatrix). Show there is a fixed point at p=0, q=0 (hyperbolic and clearly unstable).

Example 6.4. Case IV. $U(q) = cubic$. The Cubic Potential Barrier, phase space solution constructed from cases I-III:

Exercise 6.4. Show the phase space map for Hamiltonian with potential $U(q) = cubic$ (along with plot of potential).

Example 6.5. Consider the Hamiltonian: $H = a|p| + b|q|$, describe all of the consistent solutions.

1st case, $a > 0, b > 0$

$$\text{Quadrants:} \quad \begin{aligned} &\text{I:} H_I = ap + bq \\ &\text{II:} H_{II} = ap - bq \\ &\text{III:} H_{III} = ap - bq \\ &\text{IV:} H_{IV} = ap + bq \end{aligned}$$

To get the dynamics use Hamiltons eqn.'s:

Consider Quadrant I: $\dot{q} = a, \dot{p} = -b$, thus $q = at + a_0, p = -bt + b_0$. So, $q = at, p = -bt + \frac{H}{a}$ this gives the flow.

2nd case, $a < 0, b < 0$

$$\text{Quadrants :} \quad \begin{aligned} H_I &= -ap - bq \\ H_{II} &= -ap + bq \\ H_{III} &= ap + bq \\ H_{IV} &= ap - bq \end{aligned}$$

$H \leq 0$ is the only consistent solution to $a < 0, b < 0$.

3rd case, $a > 0, b < 0$

$$H_I = ap - bq \qquad\qquad \frac{dp}{dq} = \frac{b}{a}, q = 0, p = \frac{H}{a}$$

131

$$H_{II} = ap + bq \qquad\qquad \dot{q} = a, \dot{p} = b$$
$$H_{III} = -ap + bq \qquad\qquad q = at, p = bt + \frac{H}{a}$$
$$H_{IV} = ap + bq \qquad\qquad \dot{q} = -a, \dot{p} = -b \;\rightarrow\; q =$$
$$-at, p = -bt - \frac{H}{a}$$

4th case, $a < 0, b > 0$

$$H_I = -ap + bq \qquad\qquad p = 0, q = \frac{H}{b}$$
$$H_{II} = -ap - bq \qquad\qquad \dot{q} = a, \dot{p} = -b$$
$$H_{III} = ap - bq \qquad\qquad q = at + a_0, p = bt + b_0$$
$$\text{where } a_0 = 0 \qquad b_0 = \frac{H}{b}$$
$$H_{IV} = ap + bq \qquad\qquad \text{similar}$$

Exercise 6.5. What happens at $(0, 0)$?

Example 6.6. Consider potential for 1D motion with $V = -Ax^4$, $A > 0$.
$$H(x, P_x) = \frac{P_x^2}{2m} + V(x)$$

$$2mE = P_x^2 - 2mAx^4 = \left(P_x - \sqrt{2mA}x^2\right)\left(P_x + \sqrt{2mA}x^2\right)$$

There is one fixed point at the origin, $x = P_x = 0$, and the energy contoures consist of the parabolas $P_x = \pm\sqrt{2mA}x^2$ through that fixed point. The separatrix is the unstable trajectory that goes through an unstable fixed point. Have:

$$\dot{x} = \frac{\partial H}{\partial P_x} = \frac{P_x}{m} = \frac{\sqrt{2mA}x^2}{m} = \sqrt{\frac{2A}{m}}x^2$$

$$t = \frac{1}{x\sqrt{\frac{2A}{m}}} \quad \text{as } x \to 0 \ \text{and} \ t \to \infty \ \text{motion terminates.}$$

Thus, the motion terminates.

Exercise 6.6. What happens when $sqn(P_0X_0) = 1$? Show the potential and phase plots.

6.3 Review of ODE's and classification of fixed points at local, linearized (separable) level

Let's begin by shifting the origin in the phase diagram to be at a fixed point of interest and explicitly write the velocity function in terms of an expansion in position function:

$$v(r) = Ar + O(|r|^2),$$

(Eqn. 6-7)

since $v(0) = 0$ at fixed point, where A is a non-singular real matrix. Following the notation of Percival [28], let

$$A = \begin{pmatrix} a & b \\ c & d \end{pmatrix}.$$

(Eqn. 6-8)

For sufficiently small $r(x, y)$ we get only the linear term and $\dot{r} = Ar$. We would like to diagonalize the matrix A, and from there have a standardized evaluation of the fixed point behavior. To accomplish this, consider the transformation to new coordinates $R(X, Y) = Mr \rightarrow \dot{R} = BR$, where $B = MAM^{-1}$. Three cases result:

Case (1) the eigenvalues of B are real and distinct, in which case $\dot{X} = \lambda_1 X$, $\dot{Y} = \lambda_2 Y$, so

$$\left(\frac{X}{X_0}\right)^{\lambda_2} = \left(\frac{Y}{Y_0}\right)^{\lambda_1}.$$

(Eqn. 6-9)

If we have $\lambda_1 < \lambda_2 < 0$ then we have a stable node, likewise $\lambda_2 < \lambda_1 < 0$. If we have $\lambda_1 > \lambda_2 > 0$, then we have an unstable node, likewise for $\lambda_2 > \lambda_1 > 0$. If we have $\lambda_1 < 0 < \lambda_2$ we have an unstable node (a hyperbolic point); and similarly but with arrows reversed if $\lambda_2 < 0 < \lambda_1$.

Case (2) the eigenvalues of B are real and equal. There are two sub-cases: suppose $b = c = 0$, then must have $\lambda_1 = \lambda_2 < 0$ ($b = c = 0$) known as the stable star. Likewise, the $\lambda_1 = \lambda_2 > 0$ ($b = c = 0$) case is the unstable star. If, on the other hand, $c \neq 0$, then have

$$B = \begin{pmatrix} \lambda & 0 \\ c & \lambda \end{pmatrix},$$

(Eqn. 6-10)

with solution:

$$\frac{Y}{X} = \frac{c}{\lambda} \ln\left(\frac{X}{X_0}\right)$$

(Eqn. 6-11)

133

The phase curves for this case describe an improper node that is stable if $\lambda_1 = \lambda_2 < 0$ ($b \neq 0$ $c \neq 0$), or an unstable improper node if $\lambda_1 = \lambda_2 > 0$ ($b \neq 0$ $c \neq 0$).

Case (3), the eigenvalues of B are complex and conjugate to each other $\lambda_1 = \alpha + i\omega = \lambda_2$ *. Suppose the eigenvalues are pure imaginary ($\alpha = 0$), this gives rise to an elliptic point, with rotation clockwise or counterclockwise according to the sign of ω. Suppose $\alpha < 0$, we then have a stable spiral point, with rotation according to sign of ω. Likewise, if $\alpha > 0$, we then have an unstable spiral point, with rotation according to sign of ω.

Thus far we've identified the different fixed point behaviors. For first order systems all motion tends to either a fixed point or infinity, so we have a complete 'taxonomy' with what has been described thus far. For second order and higher systems this is not necessarily the case. The explicit example of the limit cycle is given next, with strange attractors left to a later section where we discuss the transition to chaos.

In our identification of fixed point behavior we've overlooked the possibility of a fixed subset that is not simply a point. Even in second order systems these can occur, resulting in the classic "limit cycle" phenomenon. Consider the following explicit case given by [28] in this regard. Suppose we have a system separable in polar coordinates according to:

$$\dot{r} = ar(r - R), \quad R > 0, and \quad \dot{\theta} = \omega.$$

The circle $r = R$ is invariant, and for motion in the neighborhood of the cycle it is either a strong attractor (stable) or the reverse (e.g., unstable, with flow lines reversed).

$$\dot{x} = x^2 \longrightarrow \frac{dx}{dt} = x^2 \longrightarrow -x^{-1} + x_0^{-1} = t$$

$$\dot{y} = -y \longrightarrow \frac{dy}{dt} = y \longrightarrow y = y_0 e^{-t}$$

Example 6.7. Unstable spiral and stable limit cycle.
For small x_1, x_2 the system:

$$\dot{x}_1 = -x_2 + x_1 r(1 - r)$$
$$\dot{x}_2 = x_1 + x_2 r(1 - r)$$
$$r^2 = x_1{}^2 + x_2{}^2$$

134

reduces to a linear system which has a center at (0,0). Show that the non-linear system has an unstable spiral at (0,0), and a stable limit cycle at r=1.

Solution

$$\dot{x}_1 = -x_2 + x_1 r(1 - r)$$
$$\dot{x}_2 = x_1 + x_2 r(1 - r)$$
$$r^2 = x_1{}^2 + x_2{}^2$$

For (x_1, x_2) both small and thus small r $(\sim x)$, have

$$\dot{x}_1 = -x_2 \quad \longrightarrow \quad \begin{pmatrix} \dot{x}_1 \\ \dot{x}_2 \end{pmatrix} = \begin{pmatrix} 0 & -1 \\ 1 & 0 \end{pmatrix}\begin{pmatrix} x_1 \\ x_2 \end{pmatrix}$$
$$\dot{x}_2 = x_1$$
$$\lambda^2 + 1 = 0 \quad \rightarrow \quad \lambda = \pm i.$$

The latter result establishes that this is an ellipsoid point{Percival], with center at (0,0). Let's now examine the r-behavior. Start by grouping:

$$x_1 \dot{x}_1 + x_2 \dot{x}_2 = (x_1{}^2 + x_2{}^2)\gamma(1 - r) = r^2(1 - r).$$

This can be rewritten:

$$\frac{1}{2}\frac{d}{dt}(x_1{}^2 + x_2{}^2) = \frac{1}{2}\frac{d}{dt}\dot{r}^2 = r^3(1 - r) \rightarrow \frac{dr}{dt} = r^2(1 - r).$$

A limit cycle is indicated at $r = 1$. To confirm,

$$dt = \frac{dr}{r^2(1 - r)}, and\ as\ r \rightarrow 1\ we\ get\ dt = \frac{dr}{1 - r}.$$

In the neighborhood of $r = 1$:

$$t = -\ln|1 - r| \quad \rightarrow \quad r = 1 \pm \exp(-t), and\ as\ t \rightarrow \infty, r$$
$$\rightarrow 1, a\ limit\ cycle.$$

Now let's consider when r is near zero. For r near zero we have $\dot{r} \cong r^2$ and since we start with $r > 0$ we will clearly have $\dot{r} > 0$ thus it spirals outwards.

Example 6.8. Elliptic fixed point (see Percival [28], p41)

Show that the origin is an elliptic fixed point for the system:

$$\dot{x}_1 = -x_2 + x_1 r^2 \sin\left(\frac{\pi}{r}\right)$$
$$\dot{x}_2 = x_1 + x_2 r^2 \sin\left(\frac{\pi}{r}\right).$$

Further, show that:
(a) the circles r=1/n, n=1,2,..., are phase curves.
(b) the trajectories between any two consecutive circles spiral either away from or towards the origin
(c) the phase curves outside of r=1 are unbounded

Solution

135

We have an elliptic point with center $(0,0)$ if $\dot{x}_1 = -x_2$ and $\dot{x}_2 = x_1$ precisely the case as r goes to zero.

(a) When we substitute r=1/n we identify these phase curves as concentric circles:

$$\dot{x}_1 = -x_2 + x_1 \left(\frac{1}{n}\right)^2 \sin(\pi n) = -x_2$$

$$\dot{x}_2 = x_1 + x_2 \left(\frac{1}{n}\right)^2 \sin(\pi n) = x_1$$

(b) Grouping the equations to get a total derivative:

$$x_1 \left(\dot{x}_1 = -x_2 + x_1 \, r^2 \, \sin\left(\frac{\pi}{r}\right)\right)$$

$$+ x_2 \left(\dot{x}_2 = x_1 + x_2 \, r^2 \, \sin\left(\frac{\pi}{r}\right)\right)$$

$$x_1 \dot{x}_1 + x_2 \dot{x}_2 = (x_1^2 + x_2^2) r^2 \sin\left(\frac{\pi}{r}\right)$$

Thus, we have:

$$\frac{1}{2}\frac{d}{dt}(x_1^2 + x_2^2) = r^4 \sin\left(\frac{\pi}{r}\right) \quad \rightarrow \quad 2r\dot{r} = 2r^4 \sin\left(\frac{\pi}{r}\right) \quad \rightarrow \quad \dot{r}$$

$$= r^3 \sin\left(\frac{\pi}{r}\right).$$

The sign of $\dot{r}$ changes according to $\sin(\pi/r)$. If we grouped to get the second solution, we would see that group spiraling inward. Between any two consecutive circles r=1/n the sign will flip. Thus, the r=1/n curves will be limit cycles $\dot{r} < 0$ if above and $\dot{r} > 0$ if below the r=1/n limit cycle.

(c) If $r > 1$, then $\sin\left(\frac{\pi}{r}\right)$ is always positive, thus $\dot{r}$ is always positive, spiraling outward.

6.4 Linear Systems and the Propagator Formalism

Case VI above is an example of a non-autonomous system, where the velocity function is an explicit function of time. For a linear second-order system (possibly by perturbation approximation to be discussed later) we have the equations:

$$\frac{d\mathbf{r}(t)}{dt} = A(t)\mathbf{r}(t) + \mathbf{b}(t).$$

(Eqn. 6-12)

Let's take $\mathbf{b}(t) = 0$, for which a 2x2 matrix valued function exists that allows us to write:

136

$$r(t_1) = K(t_1, t_0)r(t_0),$$

(Eqn. 6-13)

where the matrix $K(t_1, t_0)$ is the propagator from t_0 to t_1. Note that the propagator satisfies the Chapman-Kolmogorov relation (occurring in information theory):

$$K(t_2, t_0) = K(t_2, t_1)K(t_1, t_0)$$

(Eqn. 6-14)

The propagator matrices in this representation need not commute. Discussion about the Chapman-Kolmogorov and the deFinetti exchangeability criterion is made in later sections (quantum variants in Book 4, Stat. Mech variants inBook 5, and information theoretic issues in Book 9).

Numerous results are conveniently accessible in the propagator formalism. To get started, let's establish a relation between known solutions and the propagator matrix, to arrive at a quick transformation to the propagator formalism. Following the discussion of [28], let's start with writing the two-element column vector as a mixture of any pair of solutions:

$$r(t) = c_1 r_1(t) + c_2 r_2(t).$$

Let's now focus on the case where, at t_0, we have $r_1(t_0) = \binom{1}{0}$ and $r_2(t_0) = \binom{0}{1}$, $c_1 = x(t_0)$ and $c_2 = y(t_0)$:

$$\begin{pmatrix} x(t_1) \\ y(t_1) \end{pmatrix} = c_1 \begin{pmatrix} x_1(t_1) \\ y_1(t_1) \end{pmatrix} + c_2 \begin{pmatrix} x_2(t_1) \\ y_2(t_1) \end{pmatrix} = c_1 \begin{pmatrix} K_{11} \\ K_{21} \end{pmatrix} + c_2 \begin{pmatrix} K_{12} \\ K_{22} \end{pmatrix},$$

where the matrix values are chosen as indicated, given the special solutions chosen at t_0, and to be consistent with the eventual propagator form that is obtained:

$$\begin{pmatrix} x(t_1) \\ y(t_1) \end{pmatrix} = \begin{pmatrix} K_{11}x(t_0) \\ K_{21}x(t_0) \end{pmatrix} + \begin{pmatrix} K_{12}y(t_0) \\ K_{22}y(t_0) \end{pmatrix} = \begin{pmatrix} K_{11}x(t_0) + K_{12}y(t_0) \\ K_{21}x(t_0) + K_{22}y(t_0) \end{pmatrix}$$
$$= \begin{pmatrix} K_{11} & K_{12} \\ K_{21} & K_{22} \end{pmatrix} \begin{pmatrix} x(t_0) \\ y(t_0) \end{pmatrix}$$

Thus,

$$r(t_1) = K(t_1, t_0)r(t_0),$$

(Eqn. 6-15)

Consider Case II above, where $U(q) = +\frac{1}{2}aq^2$ (the Linear Oscillator). The solutions were found to be:

$$q = A\cos(\omega t + \delta) \quad and \quad p = -m\omega A \sin(\omega t + \delta)$$

(Eqn. 6-16)

Let t_0 correspond to $t = 0$, we then have for solution 1:

$$r_1(t_0) = \begin{pmatrix} x(t_0) \\ y(t_0) \end{pmatrix} = \begin{pmatrix} A\cos(\delta) \\ -m\omega A \sin(\delta) \end{pmatrix},$$

(Eqn. 6-17)

where we meet the special form needed if $\delta = 0$ and $A = 1$. Similarly, for $r_2(t_0)$, we choose $\delta = 90$ and $A = 1/(-m\omega)$. Thus:

$$K(t = t_1, t_0 = 0) = \begin{pmatrix} \cos(\omega t) & (m\omega)^{-1}\sin(\omega t) \\ -m\omega \sin(\omega t) & \cos(\omega t) \end{pmatrix}$$

(Eqn. 6-18)

Notice that det $K = 1$, thus describes a mapping that is area preserving, as is necessary for Hamiltonian systems. For the K matrix we have similar stability evaluations as before for B matrix, further discussion along these lines can be found at [28].

Chapter 7. Chaos

There are many ways that chaos has been exhibited in the scientific literature (see [61], others). Chaos is easily found in many one-dimensional systems that exhibit period doubling in certain regimes, where this regime of period doubling eventually turns into a chaos regime. We will examine several such systems in what follows. Other paths to chaos, such as intermittency and crises [61], when viewed graphically, have bottleneck regions in their iterative mappings, or cyclic semi-stable regions, that would explain the appearance of chaos-like behavior. Thus, the chaos examples provided will be fairly general overall.

In Sec. 7.1 we will discuss a general path to chaos phenomenon when there is periodic motion. This is because chaos is ubiquitous and with the focus on periodic motion we have a simple mathematical foundation, via an iterative map formulation, that will allow identification of domains of chaos with ease.

Before moving on with chaos, however, let's regroup for a moment and consider what is the opposite of chaos to gain a little perspective. The most ordered system is one that is "integrable" or for which there is "integrability." Recall how we used conserved quantities, as they were identified, to reduce the complexity of the differential equations, such as with the identification of angular momentum. We can represent symmetries as conserved quantities as well (Noether's theorem). If both constants of the motion and symmetries are sufficient to have a full solution to the system equations, then we have integrability, if not, then it is non-integrable. Further discussion of on integrability can be found in [38,32,37].

An example of the criticality of integrability and non-integrability to accessing chaotic behavior is conveyed by the Swinging Atwood's Machine (Fig. 7.1) [79]:

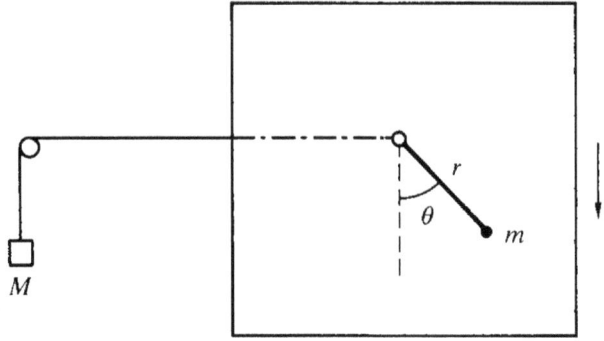

Fig. 7.1.

The Hamiltonian is

$$H = \frac{p_r^2}{2m(1+\mu)} + \frac{p_\theta^2}{2mr^2} + mgr(\mu - \cos\theta), \quad \mu = \frac{M}{m},$$

(Eqn. 7-1)

and the motion is not, in general, integrable, since H is usually the only constant of the motion.

In the case $\mu > 1$ the motion of m is always bounded by a curve of zero velocity ($p = 0$), which is an
ellipse whose shape depends on the mass ratio μ and on the energy H.

When $\mu \leq 1$ the motion is not bounded for any energy and eventually the mass M passes over the pulley.

The system is integrable in the case $\mu = 3$! In that special case, there is a second conserved quantity given by

$$J = \frac{p_\theta}{4m}\left(p_r \cos\frac{\theta}{2} - \frac{2p_\theta}{r}\sin\frac{\theta}{2}\right) + mgr^2 \sin\frac{\theta}{2}\cos^2\frac{\theta}{2}.$$

(Eqn. 7-2)

where $J = 0$. When $\mu = 3$ the motion is completely ordered. For all other mass ratios there are regions of chaotic motion.

7.1. General Path to Chaos Phenomenon: Periodic Motion → Iterative Map→ Chaos

Suppose a linear system under study, $dr(t)/dt = A(t)r(t)$ with appropriate choice of time, has parameters that are periodic in time: $A(t + T) = A(t)$ for all t. If we consider the propagator through one such period T, we have, with convenient choice for origin of time, the propagator $K = K(T, 0) = .K(nT, (n-1)T)$ Now consider the

140

propagator for nT steps in time (and use the Chapman-Kolmogorov relation) to get:

$$K(nT, 0) = K^n.$$

(Eqn. 7-3)

From the above equation we can see that systems with time-dependent parameters that are periodic in time, the propagator, $K(t, 0)$, has the property that it can be determined at certain later times, nT, merely by repeated propagations by the period propagator K. Considering that the period propagator is a linear map (and area preserving for Hamiltonian systems), this indicates that much of the future behavior (stable or not) of a periodic-parameter system can be determined by the classes of behavior under repeated period propagator mappings. In other words, the behavior of the system is mostly reduced to analysis of the behavior of its period propagation iterated map.

Let's now consider the formal definition of a "map" in the sense of a system with discrete time. The discrete-time could be due to the definition of the data (a sequence of annual readings), or due to periodicity (with measurement taken with period sampling), or for a variety of other reasons. Let's describe the system with one real-valued vector $r(t)$, now with n components, and for the discrete-time with map scenario, we suppose that $r(t + 1) = F(r(t), t)$, where F is the map function (a vector-valued function) of the phase space onto itself. For map functions that are not explicitly time-dependent we get the notation $r_{t+1} = F(r_t)$. Thus, the map formalism is very natural to the linear differential equations when there are periodic velocity functions (e.g., $dr(t)/dt = A(t)r(t)$ with $(t + T) = A(t)$). The condition of a periodic velocity function seems very powerful in this regard, and if we relax the condition of linearity we find that the iterative map result still holds.

Consider $dr(t)/dt = v(r, t)$ with $v(r, t + T) = v(r, t)$ in general (non-linear). At the first discrete time-step, t=1, we have $r(1) = F(r(0))$ by the definition of the map introduced. We then see that $dr(t + 1)/dt = v(r(t + 1), t)$, thus $r(2) = F(r(1))$ with the same mapping function, and by induction must have $r_{t+1} = F(r_t)$ in general. In other words, both autonomous and non-autonomous systems, if they have periodic velocity functions, can be described in terms of a mapping function associated with an autonomous system with discrete time. This leads to a two-step process for solving differential equations: (1) Determine the mapping function F from examination of the solution during one period of motion (from t=0 to t=1); (2) Determine solution behavior by repeated application of the mapping function. From this we see that chaotic system behavior is ubiquitous. Even simple Hamiltonian systems with one degree of freedom can exhibit chaos, or simple *conservative* Hamiltonian systems of 2 or more degrees of freedom. In fact, for systems with bounded motion, a significant portion of the phase space involves phase points that undergo chaotic motion.

In the example of the forced damped pendulum to be described next (a simple Hamiltonian system), we will find chaotic motion in a general set of circumstances. In other words, we will see that chaotic behavior (to be defined precisely) is a 'normal' outcome when pushing the perturbative limits of a system, or even if well within a perturbative domain if the parameter space pushes the 'chaos-phase' of the system. The latter description of a 'phase' of chaos in a given parameter is accurate since the parameter that enters a chaos phase (classical but indeterministic motion) for the system may exit that chaos phase, back to a domain of classical deterministic motion (and back and forth). This latter behavior is universal in first and second order systems [19], describing a set of universal parameters for classical systems at the "edge of chaos". In [45] we will see that maximal emanation/propagation of information is at the edge of chaos.

7.2 Chaos and the damped driven pendulum
Previously, for small oscillations, the pendulum oscillator was approximated as the classic spring oscillator (linear restoring force), where the differential equation describing forced oscillation with damping was (real form):

$$\ddot{x} + 2\lambda\dot{x} + \omega^2 x = \left(\frac{F}{m}\right)\cos\gamma t,$$

(Eqn. 7-4)

for which we found the solutions:

$$x(t) = a\exp(-\lambda t)\cos(\omega t + \alpha) + b\cos(\gamma t + \delta),$$

(Eqn. 7-5)

where

$$b = \frac{F}{m\sqrt{(\omega^2 - \gamma^2)^2 + (2\lambda\gamma)^2}}, \text{ and } \tan\delta = \frac{(2\lambda\gamma)}{(\omega^2 - \gamma^2)}.$$

(Eqn. 7-6)

If we don't use the small angle approximation to make $\sin x \cong x$, and we assume the pendulum wire is stiff (so a pendulum rod), we then have:

142

$$\ddot{x} + 2\lambda\dot{x} + \omega^2 \sin x = \left(\frac{F}{m}\right)\cos\gamma t.$$

(Eqn. 7-7)

Let's now consider this along the lines of the study done by [34]. First, let's change variables and overall normalize such that $\omega = 1$:

$$\ddot{\theta} + \frac{1}{q}\dot{\theta} + \sin\theta = \alpha\cos\gamma t.$$

(Eqn. 7-8)

Using the notation of [34] we have $\omega = \dot{\theta}$, not to be confused with the prior ω, to get three independent first-order equations:

(1) $\dot{\omega} = -\omega/q - \sin\theta + \alpha\cos\varphi$, where, q is the quality factor.
(2) $\dot{\theta} = \omega$
(3) $\dot{\varphi} = \gamma$

At this point we've met the two general conditions for there to exist solution domains that are chaotic:

(I) The system has three or more dynamical variables.
(II) The equations of motion contain nonlinear coupling terms.

For our problem, condition (II) is met with the coupling terms $\sin\theta$ and $\alpha\cos\varphi$. From [34], for the case where $q = 2$, we get the following behavior as we increase the driving amplitude α:

(1) $\alpha = 0.5$, the moderately driven pendulum, with simple pendulum type periodic behavior once its settled into a steady state (the trajectory is a limit cycle, thus asymptotically a cycle like with a simple pendulum).
(2) $\alpha = 1.07$, the pendulum with a double-looped trajectory in its phase diagram but with the oddity that its trajectory in a configuration diagram has yet to complete one loop even though swings exceeding 180 degrees can occur.
(3) $\alpha = 1.15$, the pendulum motion has no steady-state, it is chaotic, however its phase diagram indicates structure that is best revealed in terms of a Poincare section (which track position at multiples of the period of the forcing oscillation). For chaotic motion, the structure of the Poincare sections (phase space trajectories) is *self-similar*, this allows a precise fractal dimension to be determined [34] for the chaotic motion.
(4) $\alpha = 1.35$, the pendulum now completes a loop in configuration (real) space.

(5) $\alpha = 1.45$, the pendulum now completes two loops in configuration (real) space.

(6) $\alpha = 1.50$, the pendulum motion is chaotic

How to interpolate between the above observations, what is the boundary between systems with steady state and those without (chaotic). This is most easily represented in what's known as the bifurcation diagram (see Fig. 7.2). In the bifurcation diagram the instantaneous frequencies observed over a range of driving oscillations from $\alpha = 1$ to $\alpha = 1.50$ show a clear period-doubling behavior that multiplies quickly on approach to a domain of chaos (details to follow).

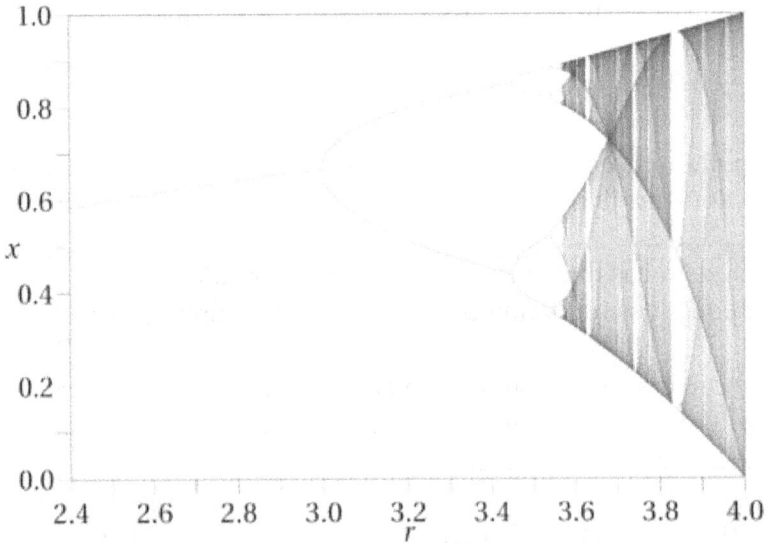

Fig. 7.2. Bifurcation Diagram for Logistic Map: $x_{n+1} = rx_n(1 - x_n)$ [80].

The bifurcation diagram most clearly captures the transition from system behavior that has steady state to behavior that is chaotic. The previous pendulum system is ubiquitous, but generating precise numerical results with it is time-consuming if all that is wanted is to demonstrate the universal behavior of chaotic systems. This is because the period-doubling transition to chaos is a distinctive trait of both second-order dynamical systems and first-order dynamical systems whose iterative mappings (Poincare Sections) involve functions of prior mapping positions that have a simple maxima [19]. General conditions for when a dynamical system with specific mapping dependency gives rise to chaotic behavior has been proven by [19] with Universal constants also thereby

144

revealed (details to follow). Rather than work with a complex evaluation at each step of the Poincare Section for, say, the pendulum, let's explore the mapping and bifurcation diagram in Fig. 7.2 that results for the much simpler logistic map, which is first-order, but whose key constants are supposedly universal so easier to evaluate this way. Here is the synopsis from [34]: "By varying the parameter r, the following behavior is observed:

- With r between 0 and 1, the population will eventually die, independent of the initial population.
- With r between 1 and 2, the population will quickly approach the value $r - 1/r$, independent of the initial population.
- With r between 2 and 3, the population will also eventually approach the same value $r - 1/r$, but first will fluctuate around that value for some time. The rate of convergence is linear, except for $r = 3$, when it is dramatically slow, less than linear (see Bifurcation memory).
- With r between 3 and $1 + \sqrt{6} \approx 3.44949$ the population will approach permanent oscillations between two values.

  These two values are dependent on r and given by   .
- With r between 3.44949 and 3.54409 (approximately), from almost all initial conditions the population will approach permanent oscillations among four values. The latter number is a root of a 12th degree polynomial (sequence A086181 in the OEIS).
- With r increasing beyond 3.54409, from almost all initial conditions the population will approach oscillations among 8 values, then 16, 32, etc. The lengths of the parameter intervals that yield oscillations of a given length decrease rapidly; the ratio between the lengths of two successive bifurcation intervals approaches the Feigenbaum constant $\delta \approx 4.66920$. This behavior is an example of a period-doubling cascade.
- At $r \approx 3.56995$ (sequence A098587 in the OEIS) is the onset of chaos, at the end of the period-doubling cascade. From almost all initial conditions, we no longer see oscillations of finite period. Slight variations in the initial population yield dramatically different results over time, a prime characteristic of chaos.

145

- Most values of r beyond 3.56995 exhibit chaotic behaviour, but there are still certain isolated ranges of r that show non-chaotic behavior; these are sometimes called *islands of stability*. For instance, beginning at $1 + \sqrt{8}$ (approximately 3.82843) there is a range of parameters r that show oscillation among three values, and for slightly higher values of r oscillation among 6 values, then 12 etc."

If the first bifurcation occurs for $\mu = \mu_1$, and the second for $\mu = \mu_2$, then it is possible to define a Universal constant F, after Feigenbaum [19]:

$$F = \lim_{k \to \infty} \frac{\mu_k - \mu_{k-1}}{\mu_{k+1} - \mu_k} = 4.66920160910299 \ldots,$$

(Eqn. 7-9)

where, remarkably, this is a universal behavior for all maps with quadratic maximum. So, in other words, for a simple (real) quadratic map or complex quadratic map (generator of Mandelbroit Set [35]) we arrive at precisely the same constant from their bifurcation maps based on the parameterization of their bifurcation events. Similarly:

Quadratic Maximum Map: $x_{n+1} = a - x_n^2$ has $\lim_{k \to \infty} \frac{a_k - a_{k-1}}{a_{k+1} - a_k} = F$.

Complex Quadratic Maximum Map Mandelbroit): $z_{n+1} = c + z_n^2$ has $\lim_{k \to \infty} \frac{c_k - c_{k-1}}{c_{k+1} - c_k} = F$.

7.3 The Special Value C_∞

For the Complex Quadratic Map the actual asymptote for the c value at the "edge of chaos" is referred to as C_∞ and has the value $C_\infty = -1.401155189 \ldots$. The constant $|C_\infty| = 1.401155189 \ldots$ is also known as Myrberg's constant [36]. The Myrberg constant, simply called C_∞ here and in [45], will play an important role in discussions.

Example 7.1. Let's consider another 1D map that is continuously differentiable with a single maximum on the interval $(0,1)$: $f(x) = \left(\frac{A}{\pi}\right) \sin \pi x$, so that we have the iterative relationship:

$$x_{n+1} = \left(\frac{A}{\pi}\right) \sin \pi x_n$$

(Eqn. 7-10)

At the first bifurcation point we have

146

$$x_{n+2} = \left(\frac{A}{\pi}\right) \sin \pi \left(\left(\frac{A}{\pi}\right) \sin \pi x_n\right) = x_n$$

Let's sketch a plot of the bifurcation diagram revealed by computational results:

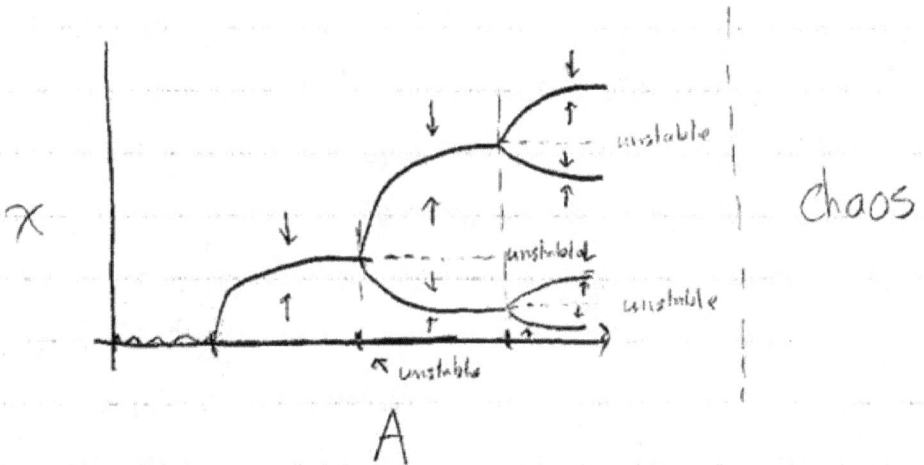

The values of A where there are the bifurcations indicated are:
$a_0 = 1$
$a_1 = 2.253804$
$a_2 = 2.614598$
$a_3 = 2.696126$
$a_4 = 2.714118$
$a_5 = 2.718112$
The Feigenbaum Number:
$$F = \lim_{j \to \infty} \frac{a_j - a_{j-1}}{a_{j+1} - a_j} \cong \frac{a_4 - a_3}{a_5 - a_4} = 4.505$$

(Eqn. 7-11)

Exercise 7.1. Redo the above analysis for another 1D map that is continuously differentiable with a single maximum on the interval (0,1).

Example 7.2. Using analytical methods, evaluate he period 1,2,... fixed points of the standard map:
$$R \to R + \varepsilon \sin \theta$$
$$\theta = \theta + R + \varepsilon \sin \theta$$
Consider the period 1 fixed points where the mapping indicates
$$R_1 = R_0 + \varepsilon \sin \theta_0 \quad \text{and} \quad \theta_1 = R_0 + \theta_0 + \varepsilon \sin \theta_0$$
while the 1-period indicates: $R_1 = R_0$ and $\theta_1 = \theta_0$, with angular equality up to a difference of $2m\pi$. Thus,
$$\sin \theta_0 = 0 \to \theta_0 = n\pi, \quad n = 0,1,2, \ldots.$$

Note that for any solution $\theta_0 = n\pi$ in the sine function there is still the solution $\theta_0 = n\pi + 2m\pi$ from multivaluedness. This is useful to recall in considering solutions to $\theta_1 = R_0 + \theta_0$:

$$R_0 = 2n\pi,$$

(not simply $R_0 = 0$). Thus, the fixed points at period 1 are: $\{\theta_0 = n\pi, R_0 = 2n\pi\}$.

Let's now consider period 2 fixed points:

$$R_2 = R_1 + \varepsilon sin\theta_1 = R_0 + \varepsilon sin\theta_0 + \varepsilon \sin(R_0 + \theta_0 + \varepsilon sin\theta_0)$$
$$\theta_2 = R_1 + \theta_1 + \varepsilon sin\theta_1$$
$$= 2(R_0 + \varepsilon sin\theta_0) + \theta_0 + \varepsilon \sin(R_0 + \theta_0 + \varepsilon sin\theta_0)$$

$R_2 = R_0 \quad \rightarrow \quad sin\theta_0 + \sin(R_0 + \theta_0 + \varepsilon sin\theta_0) = 0 \quad \rightarrow \quad \theta_0 = n\pi$ $\quad$ and $\quad R_0 = n\pi \quad$ or $\quad R_0 = 2n\pi$

$\theta_2 = \theta_0 \quad \rightarrow \quad 2(R_0 + \varepsilon sin\theta_0) + \varepsilon \sin(R_0 + \theta_0 + \varepsilon sin\theta_0) = 0 \quad \rightarrow \quad R_0$ $= n\pi \quad$ indicated.

Thus, the fixed points at period 2 are: $\{\theta_0 = n\pi, \ R_0 = n\pi\}$.

Let's now consider period 3 fixed points:

$$R_3 = R_2 + \varepsilon sin\theta_2$$
$$= R_0 + \varepsilon sin\theta_2 + \varepsilon sin(R_0 + \theta_0 + \varepsilon sin\theta_0)$$
$$+ \varepsilon sin[2R_0 + \theta_0 + \varepsilon \sin(R_0 + \theta_0)]$$

Once again we have $\theta_0 = n\pi$.

$$\theta_3 = R_2 + \theta_2 + \varepsilon sin\theta_2$$
$$= 3(R_0 + \varepsilon sin\theta_0) + 2\varepsilon \sin(R_0 + \theta_0 + \varepsilon sin\theta_0) + \theta_0$$
$$+ \varepsilon sin[2(R_0 + \varepsilon sin\theta_0) + \theta_0 + \varepsilon \sin(R_0 + \theta_0)]$$

$\theta_3 = \theta_0$:

$$0 = 3R_0 + 2\varepsilon \sin(R_0 + \theta_0) + \varepsilon sin[2R_0 + \theta_0 + \varepsilon sin(R_0 + \theta_0)].$$

Thus, the fixed points at period 3 are: $\{\theta_0 = n\pi, \ R_0 = 2n\pi\}$, and now the pattern is apparent:

Even periods have fixed points at: $\{\theta_0 = n\pi, R_0 = n\pi\}$.
Odd periods have fixed points at: $\{\theta_0 = n\pi, R_0 = 2n\pi\}$.

Exercise 7.2. Try

$$R \rightarrow R + \varepsilon[x(1 - x)]$$
$$x = x + R + \varepsilon[x(1 - x)]$$

148

Chapter 8. Canonical coordinate transformations

Previously we showed that an infinitesimal motion of an object in terms of the generalized coordinates, going from (q_0, p_0) to (q_1, p_1) in phase space, could be described in terms of the system Hamiltonian. The coordinate transformation induced by the Hamiltonian is "canonical" since its Jacobian is 1 (the area-preserving property of canonical transformations):

$$\frac{\partial(q_1, p_1)}{\partial(q_0, p_0)} = 1$$

(Eqn. 8-1)

Let's now consider the general class of such canonical coordinate transformations. Let the initial coordinates be $\{q_a, p_a\}$ for $a = 1, 2, \ldots, n$. Let the transformed coordinates be $\{Q_a, P_a\}$ (where $a = 1, 2, \ldots, n$), and we have the transformation relations:

$$q_a = q_a(\{Q_a, P_a\}; t) \ and \ p_a = p_a(\{Q_a, P_a\}; t)$$

(Eqn. 8-2)

How general an expression can we obtain for the new coordinates $\{Q_a, P_a\}$? To start. let's write Hamilton's Principle from before (with subscripts suppressed):

$$S(q, \dot{q}) = \int_{t_1}^{t_2} L(q, \dot{q}, t) dt \ ; \quad \delta S$$

$$= \left[\frac{\partial L}{\partial \dot{q}} \delta q\right]_{t_1}^{t_2} + \int_{t_1}^{t_2} \left[\left(\frac{\partial L}{\partial q}\right) - \frac{d}{dt}\left(\frac{\partial L}{\partial \dot{q}}\right)\right] \delta q dt$$

in terms of the Hamiltonian and the Action in a Modified Hamiltonian Principle (with subscripts expressed):

$$S(q_a, p_a) = \int_{t_1}^{t_2} \sum_a p_a \dot{q}_a - H(q_a, p_a, t) dt \ ; \quad \delta S$$

$$= \int_{t_1}^{t_2} \left[\sum_a \delta p_a \dot{q}_a + p_a \delta \dot{q}_a - \delta H(q_a, p_a, t) \right] dt$$

As with the Lagrangian, total time derivatives make no contribution due to the fixed endpoints (relaxing this condition is explored later). Thus, the variation in the action can be rewritten:

$$\delta S = \int_{t_1}^{t_2} \left[\sum_a \delta p_a [\dot{q}_a - \frac{\partial H}{\partial p_a}] + \delta q_a [-\dot{p}_a - \frac{\partial H}{\partial q_a}] \right] dt$$

(Eqn. 8-3)

which gives rise to Hamilton's equations when $\delta S = 0$:

$$\dot{q}_a = \frac{\partial H}{\partial p_a} \quad and \quad \dot{p}_a = -\frac{\partial H}{\partial q_a}.$$

(Eqn. 8-4)

Thus, at a guess, to retain Hamilton's equations of motion in the new variables we need to be able to express

$$\sum_a p_a \dot{q}_a - H(q_a, p_a, t)$$

$$= \sum_a P_a \dot{Q}_a - \tilde{H}(Q_a, P_a, t) + \{total\ time\ derivative\}$$

(Eqn. 8-5)

In [25] the four types of total time derivative generator functions of canonical transformations are described, with dependency on the old and new canonical variables according to {q.Q}, {q,P}, {p,Q}. {p,P} (the same generating function need not be used for all of the variables, giving rise to a mixed analysis much like the Routhian analysis involves some variables being described in terms of a Lagrangian and others in terms of a Hamiltonian). Recounting the various cases is done in detail in [25] so won't be done here. To take a specific case, consider the transform generator function of type {q.Q} and let's analyze the canonical transforms it can produce (following the conventions of [29]). Specifically, variation on:

$$\sum_a P_a \dot{Q}_a - \tilde{H}(Q_a, P_a, t) + \frac{d}{dt} F(q_a, Q_a, t),$$

(Eqn. 8-6)

which yields Hamilton's equation for the new variables as expected:

$$\dot{Q}_a = \frac{\partial \tilde{H}}{\partial P_a} \quad and \quad \dot{P}_a = -\frac{\partial \tilde{H}}{\partial Q_a}.$$

(Eqn. 8-7)

If we now take the various partial derivatives to rewrite the total time derivative, we can arrive at consistency with the Hamiltonian equations above if:

$$p_a = \frac{\partial}{\partial q_a} F(q_a, Q_a, t),$$

$$P_a = -\frac{\partial}{\partial Q_a} F(q_a, Q_a, t), \quad \tilde{H}(Q_a, P_a, t)$$

$$= H(q_a, p_a, t) + \frac{\partial}{\partial t} F(q_a, Q_a, t)$$

(Eqn. 8-8)

Thus, the Action description in a Modified Hamiltonian Principle affords a remarkable flexibility in choice of equivalent representations of the motion. The simplest thing to choose is a situation where the new coordinates are cyclic ($\dot{Q}_a = 0$ and $\dot{P}_a = 0$), and this is what is done in Hamilton-Jacobi Theory described in the next section.

8.1 The Hamiltonian-Jacobi Equation

Using the derivation and notation of [29] there is now a simple way to arrive at what is known as Hamilton-Jacobi Theory. The idea is to have a transformation such that the coordinates are cyclic. Before embarking on the canonical transformation, however, it helps to shift from function $F(q_a, Q_a, t)$ to a new function, denoted $S(q_a, P_a, t)$, by way of a Legendre transform. This new function for the condition of cyclic coordinates will be the Action as denoted by S previously. So first consider the Legendre transformation (works here since all surface terms are zero due to fixed boundary conditions):

$$F(q_a, Q_a, t) = -\sum_a P_a Q_a + S(q_a, P_a, t)$$

(Eqn. 8-9)

First the differential is by definition in terms of its dependent variables:

$$dF = \sum_a \left(\frac{\partial F}{\partial q_a} dq_a + \frac{\partial F}{\partial Q_a} dQ_a\right) + \frac{\partial F}{\partial t} dt$$

$$= \sum_a (p_a dq_a - P_a dQ_a) + \frac{\partial F}{\partial t} dt$$

but from above also have:

151

$$dF = -\sum_a (P_a dQ_a + dP_a Q_a) + dS$$

(Eqn. 8-10)

Thus,

$$dS = \sum_a (p_a dq_a + Q_a dP_a) + \frac{\partial F}{\partial t} dt,$$

(Eqn. 8-11)

where we can see that the functional dependence is indeed $S(q_a, P_a, t)$. If we take the following relations by definition for partial derivative for:

$$p_a = \frac{\partial}{\partial q_a} S(q_a, P_a, t),$$

$$Q_a = \frac{\partial}{\partial P_a} S(q_a, P_a, t), \qquad \frac{\partial}{\partial t} S(q_a, P_a, t) = \frac{\partial}{\partial t} F(q_a, Q_a, t)$$

(Eqn. 8-12)

we then get:

$$\tilde{H}(Q_a, P_a, t) = H(q_a, p_a, t) + \frac{\partial}{\partial t} S(q_a, P_a, t)$$

(Eqn. 8-13)

Any $S(q_a, P_a, t)$ given the above partials will generate a canonical transformation by construction. Let's now choose a canonical transformation with $S(q_a, P_a, t)$ such that $\tilde{H}(Q_a, P_a, t) = 0$, since $\tilde{H}$ thereby has no dependence on Q_a and P_a they are cyclic coordinates. In which case we arrive at:

$$0 = H(q_a, p_a, t) + \frac{\partial}{\partial t} S(q_a, P_a, t) = H\left(q_a, \frac{\partial S}{\partial q_a}, t\right) + \frac{\partial}{\partial t} S(q_a, P_a, t)$$

and since Q_a and P_a are constants of the motion, we then get the Hamilton-Jacobi Equation:

$$H\left(q_a, \frac{\partial S}{\partial q_a}, t\right) + \frac{\partial}{\partial t} S(q_a, t) = 0$$

(Eqn. 8-14)

This is a first-order partial differential equation that can be solved by introducing (n+1) constants of integration ($\{c_a\}$ and S_0):

$$S = S(q_a, c_a, t) + S_0$$

If we choose the constants $\{c_a\}$ to be the constants $\{P_a\}$ we return to the classic form of the solution known as Hamilton's Principle Function:

$$S = S(q_a, P_a, t) + S_0$$

(Eqn. 8-15)

where

$$p_a = \frac{\partial}{\partial q_a} S(q_a, P_a, t), \qquad Q_a = \frac{\partial}{\partial P_a} S(q_a, P_a, t).$$

(Eqn. 8-16)

The reason this form is significant is due to the latter relation given that $\{P_a\}$ and $\{Q_a\}$ are constants of the motion it is invertible to give a description of the motion that is only a function of time:

$$q_a = q_a(\{Q_a\}, \{P_a\}, t)$$

Thus, the motion is clearly defined as a path (parameterized by t). Let's consider the derivative of S along this path:

$$\frac{dS}{dt} = \sum_a \frac{\partial S}{\partial q_a} \dot{q}_a + \frac{\partial S}{\partial t} = \sum_a p_a \dot{q}_a - H = L(q_a, \dot{q}_a, t)$$

Thus,

$$S = \int_{t_0}^{t} L(q_a, \dot{q}_a, \tau) d\tau + S_0(t_0)$$

(Eqn. 8-17)

Or, changing the time variable notation slightly, we arrive at the form originally posited as Hamilton's "action formulation" mentioned at the start of Ch. 3:

$$S = \int_{t_1}^{t_2} L(q, \dot{q}, t) dt$$

(Eqn. 8-18)

Example 8.1. Let's start with an expression for the action:

$$S = (q, q_0, t, t_0) = \frac{m\omega}{2\sin\omega t} \{(q^2 + q_0{}^2)\cos\omega t - 2qq_0\}; \qquad T = t - t_0.$$

What system results? What is the Hamiltonian? What are the trajectories?

Solution:

$$H = -\frac{\partial S}{\partial t} = \frac{m\omega^2}{(2\sin\omega t)^2} \{-4qq_0\cos\omega t + 2(q^2 + q_0{}^2)\}.$$

From which we can reconstruct

153

$$p = \frac{\partial S}{\partial q} = \frac{m\omega}{2sin\omega t}\{2qcos\omega t - 2q_0\}$$

$$p^2 = 2m\left[\frac{m\omega^2}{2sin^2\omega t}\right][q^2cos^2\omega t - 2qq_0cos\omega t + q_0{}^2]$$

$$\frac{p^2}{2m} = \frac{m\omega^2}{(2sin\omega t)^2}\{-2q^2sin^2\omega t - 4qq_0cos\omega t + 2(q^2 + q_0{}^2)\}.$$

Thus, the Hamiltonian can be written as:

$$H = \frac{p^2}{2m} + \frac{m\omega^2}{(2sin\omega t)^2}\{2q^2sin^2\omega t\} = \frac{p^2}{2m} + \frac{m\omega^2 q^2}{2} = \frac{1}{2m}[p^2 + m^2\omega^2 q^2].$$

Thus the conserved quantity, energy, is:

$$E = \frac{1}{2m}[p^2 + m^2\omega^2 q^2].$$

This is a harmonic oscillator. Let's get the trajectories now:

$$\dot{q} = \frac{\partial H}{\partial p} = \frac{p}{m} \quad and \quad \dot{p} = -\frac{\partial H}{\partial q} = m\omega^2 q.$$

One possible set of solutions:

$$q = \sqrt{2E/m\omega^2}cos\omega t \quad and \quad p = \sqrt{2mE}sin\omega t.$$

Exercise 8.1. Find all solutions.

Example 8.2. Solve the H-J equation for motion in one dimension for a particle is acted on by a force that is constant in both space and time.

Solution

The H-J equation in 1D:

$$H(q,p) + \frac{\partial S}{\partial t} = 0, \quad with \quad p = \frac{\partial S}{\partial q}, \quad thus \quad H\left(q, \frac{\partial S}{\partial q}\right) + \frac{\partial S}{\partial t} = 0.$$

(a) For particle in 1D, nonrelativistic, with force constant in space and time, have:

$$F = -\frac{\partial V}{\partial q} = \alpha \quad \rightarrow \quad V = -\alpha q,$$

and for the kinetic energy we have the usual:

$$T = \frac{1}{2}m\dot{q}^2.$$

The Lagrangian is thus:

$$L = T - V = \frac{1}{2}m\dot{q}^2 + \alpha q.$$

Now to construct the Hamiltonian, first the momentum:

$$p = \frac{\partial L}{\partial \dot{q}} = m\dot{q},$$

Thus:

$$H(q, p, t) = \dot{q}p - L = \frac{p^2}{m} - \frac{1}{2}m\left(\frac{p}{m}\right)^2 - \alpha q = \frac{p^2}{2m} - \alpha q.$$

Using this in the 1D H-J equation, we get:

$$\frac{1}{2m}\left(\frac{\partial S}{\partial q}\right)^2 + \alpha q + \frac{\partial S}{\partial t} = 0.$$

If we guess a solution of the form:

$$S(q, E, t) = w(q, E) - Et \longrightarrow \frac{\partial S}{\partial t} + H = 0 \longrightarrow H = E.$$

Solving for the function $w(q, E)$:

$$\frac{1}{2m}\left(\frac{\partial w}{\partial q}\right)^2 = E - \alpha q \longrightarrow \frac{\partial w}{\partial q} = \sqrt{2m(E - \alpha q)}.$$

Thus,

$$S = \sqrt{2mE}\int dq\sqrt{1 - \frac{\alpha q}{E}} - Et \longrightarrow S$$

$$= \sqrt{2mE} \cdot \frac{2\sqrt{\left(1 - \frac{\alpha q}{E}\right)^3}}{3\left(-\frac{\alpha}{E}\right)} - Et + f(x_0)$$

Exercise 8.2. Solve the H-J equation for motion in one dimension for a particle is acted on by a force that is constant in space and increasing linearly in time.

8.2 From Hamilton-Jacobi equation to Schrodinger Equation

Classical mechanics thus far has been non-relativistic and non-field, except in an idealized sense for the latter. Furthermore, when matter accretes gravitationally we understand its collapse to be halted at some point by material compression properties that themselves trace to electrodynamics non-collapse solutions. So our objects thus far have been simplified to their classical non-electrodynamic behavior. Once we attempt to account for relativity or describe fields as dynamical in their own right we encounter new complications (such as electrodynamics radiative collapse) and a quantum theory is indicated. There are three main formalisms that connect the classical theory to a quantum theory (Schrodinger, Heisenberg, and Feynman-Dirac). There's also the older Bohr-Sommerfeld Quantization in an earlier attempt comprising a semiclassical solution in the current theory. The first to be discussed is the Schrodinger wave-equation form of quantization, which is directly related to the Hamilton-Jacobi equation with appropriate substitution of operators.

The classical Hamilton-Jacobi equation has the differential $\partial/\partial q_a$:

$$H\left(q_a, \frac{\partial S}{\partial q_a}, t\right) + \frac{\partial}{\partial t} S(q_a, t) = 0$$

(Eqn. 8-19)

In the Schrodinger quantum theory we switch to a wave-function operator formalism, which begins with a wave-function of the form:

$$\psi(q_a, t) \propto e^{\frac{i}{\hbar}S(q_a,t)},$$

(Eqn. 8-20)

where we see the action entering as a phase in the wave-function. Acting on the wave-function is an operator expression whereby p_a is not substituted by $\frac{\partial S}{\partial q_a}$ (classical expression) but by $\frac{\partial}{\partial q_a}$ as part of an operator expression:

$$H(q_a, p_a, t) + \frac{\partial}{\partial t} S(q_a, t) = 0 \rightarrow \left\{H\left(q_a, \frac{\partial}{\partial q_a}, t\right) + \frac{\partial}{\partial t}\right\} \exp \frac{i}{\hbar} S(q_a, t)$$
$$= 0$$

(Eqn. 8-21)

the latter being a form of Schrodinger's equation (further details in [42]). The quantum equation of motion, to first order in $\frac{S}{\hbar}$, then recovers classical mechanics, since

$$\left\{H\left(q_a, \frac{\partial S}{\partial q_a}, t\right) + \frac{\partial S}{\partial t}\right\} \exp \frac{i}{\hbar} S(q_a, t) = 0 \rightarrow H\left(q_a, \frac{\partial S}{\partial q_a}, t\right) + \frac{\partial}{\partial t} S(q_a, t)$$
$$= 0.$$

(Eqn. 8-22)

Semiclassical physics then describes the initial mix of second and higher order terms that give rise to non-classical effects.

For bounded configurations full solutions to Schrodinger's equations are possible, such as for the critical Hydrogen atom. When applied to the hydrogen atom, quantum physics solves a conundrum of classical electrostatics whereby the hydrogen atom has stable bound states (and doesn't simply collapse).

Example 8.3. Consider the time-dependent Schrodinger equation for a single particle in a potential $U(r, t)$. This quantum mechanical problem will be studied extensively in [42], but viewed in a general sense now it is very instructive as to the new "place" that awaits for classical mechanics in the larger quantum mechanical world). Consider the ansatz where the wavefunction solution can be written:

156

$$\Psi(r,t) = A(r,t) \exp\left[\frac{i}{\hbar}\theta(r,t)\right],$$

(Eqn. 8-23)

where A and θ are real and analytic in $\hbar$. (a) Show the expansion in $\hbar$ leads, to lowest order, to θ being a solution to the corresponding H-J equation (it is the classical Action). (b) Show at next order in $\hbar$ that A^2 satisfies an equation of continuity (this will help motivate the Born interpretation in [42]).

Solution

(a) We have for the time-dependent Schrodinger equation:

$$i\hbar\frac{\partial}{\partial t}\Psi(r,t) = \hat{H}\Psi(r,t).$$

For a single particle in a potential we have:

$$\hat{H} = \frac{\hat{p}^2}{2m} + \hat{U}(r,t) = -\frac{\hbar^2}{2m}\nabla^2 + U(r,t),$$

thus,

$$i\hbar\frac{\partial}{\partial t}\Psi(r,t) = -\frac{\hbar^2}{2m}\nabla^2\Psi(r,t) + U(r,t)\Psi(r,t).$$

Let's now try the indicated solution to get an equation in terms of $\{A, \theta\}$:

$$i\hbar\frac{\partial A}{\partial t} - A\frac{\partial\theta}{\partial t} = -\frac{\hbar^2}{2m}\nabla^2 A - \frac{i\hbar}{m}\nabla A\nabla\theta + \frac{A}{2m}(\nabla\theta)^2 - \frac{i\hbar}{2m}A\nabla^2\theta + AU.$$

At zeroth order in $\hbar$, $\hbar^0$, we have the terms:

$$\frac{\partial\theta}{\partial t} = -\left[\frac{(\nabla\theta)^2}{2m} + U\right].$$

The H-J (Hamilton-Jacobi) equation for the θ variable is:

$$H(r,\nabla\theta) + \frac{\partial\theta}{\partial t} = 0 \rightarrow \frac{\partial\theta}{\partial t} = -\left[\frac{(\nabla\theta)^2}{2m} + U\right],$$

which is precisely the zeroth order relationship.

(b) At first order in $\hbar$, $\hbar^1$, we have the terms:

$$i\hbar\frac{\partial A}{\partial t} = -\frac{i\hbar}{m}\nabla A\nabla\theta - \frac{i\hbar}{2m}A\nabla^2\theta,$$

multiplying by A and regrouping:

$$\frac{\partial A^2}{\partial t} = -\frac{1}{m}\nabla(A^2\nabla\theta) \rightarrow \frac{\partial\rho}{\partial t} = -\nabla\left(\rho\frac{\nabla\theta}{m}\right), where \; \rho = A^2,$$

Thus, we get:

$$\frac{\partial\rho}{\partial t} + \nabla\cdot(\rho v) = 0, where \; v = \frac{\nabla\theta}{m},$$

157

where ρ is like a fluid density and v is like a flow velocity vector field.

Exercise 8.3. What is revealed at second order in $\hbar$?

8.3 Action-Angle Variables and Bohr/Sommerfeld-Wilson Quantization

For the special case of bounded conservative motion that is separable and periodic we can switch to what are known as the action-angle variables. The "action variables" are defined as the integral of the area in phase space over one period of the motion for each degree of freedom:

$$J_a = \oint p_a dq_a$$

(Eqn. 8-24)

The resulting J_a are only dependent on the constants of the motion, here denoted $\{\alpha_a\}$ and following the notation of [29]:

$$J_a = J_a(\{\alpha_a\}).$$

(Eqn. 8-25)

Or, inverting, and renaming $\alpha_1 = E$:

$$E = H(\{J_a\}).$$

(Eqn. 8-26)

Further details on the derivation can be found in [29]. From here we can determine the fundamental frequencies of the system in terms of the above Hamiltonian expressed via action variables:

$$v_a = \frac{\partial}{\partial J_a} H(\{J_a\}).$$

(Eqn. 8-27)

In Sommerfeld-Wilson quantization it was proposed that the action variables should be quantized with integer amounts of Plank's constant:

$$J_a = \oint p_a dq_a = nh$$

(Eqn. 8-28)

8.4 Poisson Brackets

Poisson Brackets take on a special form when working in canonical coordinates, and they are defined in terms of a Hamiltonian regardless, so the presentation of Poisson Brackets is placed here for that reason. In canonical coordinates let's consider two functions $f(q_i, p_i, t)$ and $g(q_i, p_i, t)$, where the canonical coordinates (on some phase space) are given by $\{p_i, q_i\}$ where $i = 1..N$. The Poisson bracket function of these two functions is denoted by $\{f, g\}$ and defined by:

158

$$\{f,g\} = \sum_{i=1}^{N} \left(\frac{\partial f}{\partial q_i} \frac{\partial g}{\partial p_i} - \frac{\partial f}{\partial p_i} \frac{\partial g}{\partial q_i} \right).$$

(Eqn. 8-29)

Thus, by definition we have:

$$\{q_i, q_j\} = 0, \quad \{p_i, p_j\} = 0, \quad and \quad \{q_i, p_j\} = \delta_{ij},$$

(Eqn. 8-30)

where the Kronecker delta is used ($\delta_{ij} = 1$ if $i = j$ and $\delta_{ij} = 0$ otherwise).

Often, we examine the time evolution of a function on the symplectic manifold induced by the one-parameter family of symplectomorphisms (canonical and area preserving diffeomorphisms) [37], where Poisson brackets are preserved.

We will see Poisson brackets again in [42] on Quantum mechanics as generalized Poisson brackets, which upon quantization deform to Moyal brackets (a generalization of the Lie algebra, the Poisson algebra, associated with the Poisson Brackets). In terms of Hilbert space, we arrive at non-zero quantum commutators.

Chapter 9. Perturbation theory, Dimensional analysis, and Phenomenology

9.1 Hamiltonian Perturbation Theory

In perturbation theory we consider a known solution, or system (typically a Hamiltonian description with its constants of the motion made clear), and we consider a small "perturbation" to that system. We then do a perturbation expansion for our solution by solving at various orders separately on what are simpler differential problems (see App. A. for some discussion and examples of ODE perturbation solution methods in general).

Example 9.1. Perturbation theory involving a full Hamiltonian.
Let's now consider perturbation theory involving a Full Hamiltonian $H(q, p, t)$, a simpler Hamiltonian with known solutions $H_0(q, p, t)$, and the perturbation part $\Delta H(q, p, t)$, where $\Delta H \ll H_0$:
$$H(q, p, t) = H_0(q, p, t) + \Delta H(q, p, t).$$
(Eqn. 9-1)
We expand all variables to various orders in a perturbation parameter (appearing in ΔH).

Consider the example of free motion with spring restoring force seen as perturbation. In this instance we know the full solution without any perturbation theory so can see how our result performs. So, for H_0 we have $H_0 = p^2/2m$ and for perturbation let's use the solution form for the spring potential in canonical coordinates: $\Delta H = (m\omega^2/2)x^2$. We can then evaluate the Hamilton equations to get the usual result:
$$\dot{x} = \frac{p}{m} \quad and \quad \dot{p} = -m\omega^2 x$$
(Eqn. 9-2)
(without any approximation). Treated as a perturbation, let's consider ω^2 as the perturbation parameter, thus at zeroth order we have $\dot{p}_0 = 0$ and $\dot{x}_0 = p_0/m$. Thus
$$p^{(0)} = p_0 = const. \quad and \quad x^{(0)} = x_0 = \left(\frac{p_0}{m}\right)t,$$
(Eqn. 9-3)
where we choose initial condition $x(t = 0) = 0$. Now, at first order we get:

$$\dot{p}^{(1)} = -m\omega^2 x^{(0)} = -\omega^2 p_0 t \quad \rightarrow \quad p^{(1)}(t) = p_0 - \frac{1}{2}\omega^2 p_0 t^2$$

(Eqn. 9-4)

and

$$\dot{x}^{(1)} = \frac{p^{(1)}}{m} = \frac{p_0}{m} - \frac{1}{2m}\omega^2 p_0 t^2 \quad \rightarrow \quad x^{(1)}(t) = \frac{p_0}{m}t - \frac{1}{6m}\omega^2 p_0 t^3.$$

(Eqn. 9-5)

If we now compare with the known full solution:

$$p(t) = p_0 \cos \omega t \quad and \quad x(t) = \frac{p_0}{m\omega}\sin \omega t,$$

(Eqn. 9-6)

thru first order we can see exact agreement.

If there is a time-dependent perturbation, then one often shifts from a Hamiltonian formulation to the Hamiltonian-Jacobi formulation [37]. Consider the $H = H_0 + \Delta H$ setup as before, but now we have the added information of having obtained the principal function S that is the generating function for the canonical transformation from $\{q, p\} \rightarrow \{\alpha, \beta\}$ such that:

$$H_0\left(q, \frac{\partial S}{\partial q}, t\right) + \frac{\partial}{\partial t}S(q, \alpha, t) = 0.$$

(Eqn. 9-7)

In relation to H_0, the variables $\{\alpha, \beta\}$ are canonical and thus constants. In relation to H they will not be constants but will still be chosen as our canonical variables (let $\{P = \alpha, Q = \beta\}$):

$$P = \alpha(q, p) \quad and \quad Q = \beta(q, p).$$

(Eqn. 9-8)

Recasting to standard H-J form for perturbed Hamiltonian H with the time-dependent perturbation:

$$H(\alpha, \beta, t) = H_0(\alpha, \beta, t) + \Delta H(\alpha, \beta, t) + \frac{\partial S}{\partial t} = \Delta H(\alpha, \beta, t),$$

(Eqn. 9-9)

and since $\dot{Q} = \frac{\partial H}{\partial P}$ and $\dot{P} = -\frac{\partial H}{\partial Q}$ we get the exact relations:

$$\dot{\alpha} = -\frac{\partial \Delta H}{\partial \beta} \quad and \quad \dot{\beta} = \frac{\partial \Delta H}{\partial \alpha}.$$

(Eqn. 9-10)

Exact solutions are often not possible, so we do perturbation expansions as before. Here, whatever values for $\{\alpha, \beta\}$ obtained at zeroth-order are then used in computing first-order, as before:

162

$$\dot{\alpha}^{(1)} = -\frac{\partial \Delta H}{\partial \beta}, \quad where \ \alpha = \alpha^{(0)} \ and \ \beta = \beta^{(0)},$$

(Eqn. 9-11)

and similarly for $\dot{\beta}^{(1)}$, and then iterated at higher order as needed.

Exercise 9.1. Apply the H-J perturbation approach to the spring system considered previously and re-obtain the result in the H-J formalism.

9.2 Dimensional analysis

Physics has dimensional quantities, unlike the differential mathematics used thus far (although one can introduce mathematical elements that can act as dimensional quantities). Dimensionless quantities can be grouped into products that are dimensionless. For Example, the Stefan-Boltzmann Law (described in [42,45]), gives a relation between radiant energy E in a cavity, of volume V, with walls at Temperature T:

$$\frac{E}{V} = \frac{8\pi^5}{15} \frac{k_B^4 T^4}{c^3 h^3}.$$

(Eqn. 9-12)

Physics mathematical formulas must have consistency on the dimensionality of terms.

Example 9.2. A marble rolling in a circular orbit

Consider a marble rolling in a circular orbit inside an inverted cone (see [62] for more such examples), with half-angle (from vertical) equal to θ. The variables for the system are then the orbital period τ, mass m, radius of orbit R, acceleration due to gravity g, and the aforementioned θ. Let's make a dimensionless product:

$$\tau^\alpha m^\beta R^\gamma g^\delta = [T]^\alpha [M]^\beta [L]^\gamma [LT^{-2}]^\delta = T^{\alpha - 2\delta} M^\beta L^{\gamma + \delta},$$

(Eqn. 9-13)

which is dimensionless if $\alpha - 2\delta = 0$ and $\beta = 0$ and $\gamma + \delta = 0$, or simplifying we get:

$$\beta = 0 \quad and \quad \gamma = -\delta = -\alpha/2.$$

Thus, we have the relation:

$$\tau = \sqrt{\frac{R}{g}} f(\theta).$$

(Eqn. 9-14)

With a lot more effort, a detailed analysis shows that $f(\theta) = 2\pi\sqrt{\tan\theta}$.

Exercise 9.2. Show that $f(\theta) = 2\pi\sqrt{\tan\theta}$.

163

A more general formulation of the partial solution possible by dimensional analysis is given by the Buckingham Π Theorem [62].

9.2.1 Buckingham Π Theorem

I. If an equation is dimensionally homogenous it can be reduced to a relationship among a complete set of independent dimensionless products [63]

II. The number of complete and independent dimensionless Products N_P is equal to the number of dimensionless Variables (and constants) N_V minus the number of Dimensions N_D needed to express the formulae: $N_P = N_V - N_D$.

Clarification of the above methods is best shown with a few examples.

Example 9.3. Pendulum dimensional analysis.
For a pendulum with period τ, mass m, arm length l, acceleration due to gravity g:
$$\tau^\alpha m^\beta l^\gamma g^\delta = [T]^\alpha [M]^\beta [L]^\gamma [LT^{-2}]^\delta = T^{\alpha-2\delta} M^\beta L^{\gamma+\delta},$$
which has the same solution as before (but with no θ), thus we have:
$$\tau = C \sqrt{\frac{l}{g}},$$
where C is a constant.

Exercise 9.3. Redo for horizontal spring motion on frictionless surface, one end attached, the other with a non-negligible mass.

Example 9.4. Nuclear Blast Analysis by G.I. Taylor [33]
This is a famous example where the yield (energy) of a nuclear explosion was determined from a sequence of high-speed photographs that were published in a Newspaper (with the necessary timestamps showing the spread of the blast). Let R denote the radius of an expanding blast wave, let time from blast be t, let the energy released be E, and let the (initial) atmospheric density be ρ.

Exercise 9.4. Show that $E = k\rho R^5/t^2$ for some (dimensionless) constant k.

Example 9.5. Consider the Hamiltonian:

$$H = \frac{1}{2}\left(P_x{}^2 + P_y{}^2\right) + 2x^3 + xy^2$$

For which the Hamiltonian equations give:

$$\dot{x} = P_x; \quad \dot{y} = P_y; \quad \dot{P}_x = -(6x^2 + y^2); \quad \dot{P}_y = -(2xy).$$

We have our first conserved quantity the Energy, $E = H$, and referring to the Energy dimensionality let's construct a table of terms:

Term	Order in E
x, y	1/3
P_x, P_y	½
$\dfrac{d}{dt}$	1/6
H	1

We want a second conserved quantity W such that $\dot{W}$ can be constructed from $(x, y, P_x, P_y, \dot{x}, \dot{y}, \dot{P}_x, \dot{P}_y)$ so as to give zero consistent with the form of the "building blocks" above. Since the $\dot{P}_x, \dot{P}_y$ are the only place where terms are coupled, they must be in W. Since the $\dot{P}_x, \dot{P}_y$ are of order 2/3, we must have $\dot{W}$ of order $\geq 2/3$. Also, W must be an exact differential (as with H).

Case 1: consider $\dot{W}$ to be order 2/3, this means that:

$$\dot{W} = \alpha \dot{P}_x + \beta \, \dot{P}_y + ax^2 + bxy + cy^2,$$

where the coefficients are all constants we can choose. This expression isn't an exact differential for any choice of constants, however, so this case doesn't work.

Case 2: consider $\dot{W}$ to be order 5/6, this means that:

$$\dot{W} = \alpha x P_x + \beta y P_x + \gamma y P_y + \delta x P_y + ax\dot{x} + bx\dot{y} + cy\dot{x} + dy\dot{y}.$$

This expression isn't an exact differential either, so this case doesn't work.

Case 3: consider $\dot{W}$ to be order 6/6, ... have terms like $x\dot{P}_x$, and again, no solution.

Case 4: consider $\dot{W}$ to be order 7/6, this works, but it recovers the first conserved quantity, the Hamiltonian itself.

Case 5: consider $\dot{W}$ to be order 8/6, ... have terms like $x^2\dot{P}_x$, and again, no solution.

Case 6: consider $\dot{W}$ to be order 9/6, ... this works. The general form is now:

$$\dot{W} \propto E^{3/2} \quad \rightarrow \quad W \propto E^{4/3}$$

The general expression for W is now:

$$W = a_1 x^4 + a_2 x^3 y + a_3 x^2 y^2 + a_4 x y^3 + a_5 y^4$$
$$+ b_1 x P_x^2 + b_2 x P_x P_y + b_3 x P_y^2 + b_4 y P_x^2 + b_5 y P_x P_y + b_6 y P_y^2$$

The general expression for $\dot{W}$ is thus:

$$\dot{W} = x^3 P_x (4a_1 - 12b_1) + \cdots,$$

where the constant coefficients for each term are each separately equal to zero. There are thus 12 equations for the 11 unknowns indicated. Solving, we find that:

$$W = x^2 y^2 + \frac{1}{4} y^4 - x P_y^2 + y P_x P_y.$$

9.2.2 Dimensional Analysis Shows 22 Unique dimensional quantities [62]

If we start with the set of 6 fundamental dimensional constants, $\{G, \varepsilon_0, c, e, m_e, h\}$, we find that there are 22 unique dimensionful groupings [62] and 2 dimensionless groupings (the Eddington-Dirac number and the fine structure constant). In [45] we will again find 22 fundamental, dimensionful, parameters are indicated.

Exercise 9.5. Identify the 22 dimensionful groupings.

9.3 Phenomenology

When you don't have a fundamental theory but still want to establish a scientific model based on some empirical data of some phenomenon, then what you are establishing is a phenomenological model. A phenomenological model is not based on any first principles. Fundamental theories often start as phenomenological models until they are better understood. Feynman in his descriptions of Physical Law [64], for example, describes the discovery process for physical law as enlightened guesswork. Thermodynamics is often viewed as a phenomenological theory that has borrowed physical law from elsewhere (such as conservation of energy). Partly for this reason, and awaiting other developments of the theory, discussion of phenomenology in the thermodynamics and statistical mechanics contexts isn't done until [44].

166

Some of the hardest problems in modern theoretical physics have been addressed in the form of phenomenological models (particle physics, condensed matter physics, plasma physics). If all else fails, try phenomenology. A famous example of this from the Film "Dark Star" has to do with deactivating a "thermostellar" bomb that has accidentally activated (it's the semi-truck shaped object shown in Fig. 8.1). The bomb is controlled by an AI and the crew has deemed their best chance to deactivate the bomb is to "teach it phenomenology," so that it can see the big picture and realize it doesn't have to explode if it doesn't want to…..
Unfortunately, upon re-evaluating with greater perspective, the AI decides it is god, says "Let there be light," and explodes. This is usually how things work out in Physics as well, but that will have to await another day and another book (see the forthcoming [40] for description of electromagnetism).

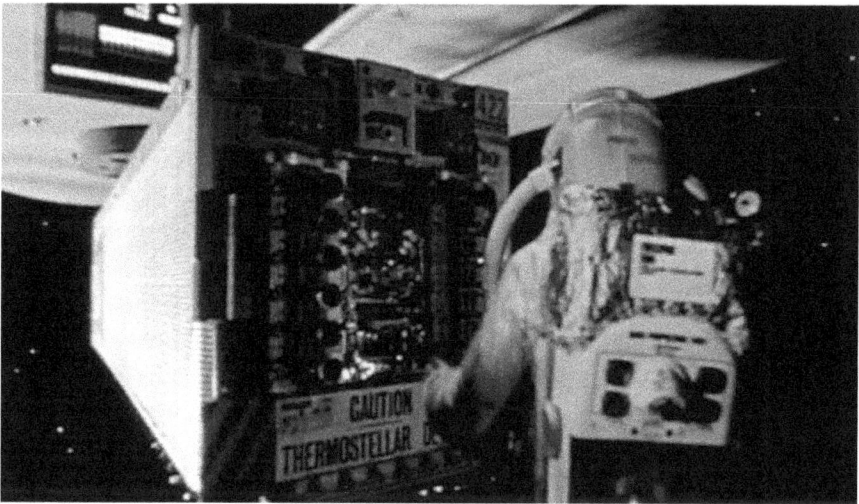

Fig. 9.1 Crewmember shown teaching the bomb's AI phenomenology, from film "Dark Star".

Chapter 10. Extra Exercises

Exercise 10.1.
Consider a collision of two identical systems, each consisting of two point masses m joined by a spring of constant k. Before the collision, each spring is "relaxed", or uncompressed. Before the collision, one system is moving at speed v toward the other, along the line of the springs and the second system is at rest. The particles that collide stick together to form a 3-particle system as shown in the "after" picture. If the collision time is short compared to $\sqrt{\frac{m}{k}}$, $find$

(a) The speed of each of the three final particles immediately after the collision.
(b) The position of the particle on the far right as a function of the time t after the collision

Exercise 10.2.
Two particles of masses m_1, m_2 and positions $\vec{r}_1$, $\vec{r}_2$, respectively, interact with potential energy $U(r)$, where $\mathrm{r} = \left| \vec{r}_1 - \vec{r}_2 \right|$.

(a) Write the Lagrangian L of this system.
(b) Define the relative coordinate $\vec{r} = \vec{r}_1 - \vec{r}_2$ and the center of mass coordinate $\vec{R} = \frac{\left(m_1 \vec{r}_1 + m_2 \vec{r}_2 \right)}{(m_1 + m_2)}$. Express the Lagrangian L in terms of these generalized coordinates. Show that $L = L_R + L_r$, where L_R is the part of the Lagrangian containing the coordinate $\vec{R}$ and L_r is the part containing the coordinate $\vec{r}$. Write L_r in the form of the Lagrangian of a single particle having coordinate $\vec{r}$ and mass m. Give the expression for this "reduced mass m in terms of m_1 and m_2.
(c) In the rest of the problem, consider the motion of the particle described by the Lagrangian L_r
(*the subscript r on L will be dropped for brevity*). Choose cylindrical coordinates with the z-axis pointing in the direction of the angular momentum $\vec{l} = \vec{r} \times \vec{p}$ where $P_i = \partial L / \partial \dot{r}_i$. Write the Lagrangian in cynlindrical coordinates (r, ϕ, z).

(d) Now show that the angular momentum is conserved. Since $\vec{l}$ is conserved, the particle can be assumed to move in the plane $z = 0$. This simplifies the Lagrangian.

(e) Show that as a result of Lagrange's equations, there is a conserved energy E, and give it explicitly in terms of r, ϕ and their time derivatives. Write the expression for the conserved angular

(f) From the expression for E express t as an integral function of r and the constants of motion E and l.

(g) Similarly, express ϕ as an integral function of $r, E,$ and l.

Exercise 10.3.

A particle if mass m moves in a force field of the form

$$\vec{F} - \left(-\frac{a}{r^2} + \frac{b}{r^{\frac{3}{2}}} \right) \hat{r}$$

Where a and b are positive constants.

(a) For what range of radil are circular orbits possible?

(b) For what range of radil are circular orbits stable?

(c) Find the frequency of small oscillations about a circular orbit of radius r $= \dfrac{a^2}{4b^2}$

Exercise 10.4.

(a) Show that an isolated particle with finite rest mass m cannot decay into a single particle with zero rest mass.

(b) Can a single particle with zero rest mass decay into n particles, all having zero rest mass and positive energy? If so, give an example. If not, prove that it is impossible for all n > 1

Exercise 10.5.

A rod of length, a, and mass, m, is suspended from a massless string of length a/3. Obtain the normal-mode frequencies (eigen frequencies) for small displacements from the stable equilibrium position of this system.

Exercise 10.6.

Consider the transverse motion (i.e motion perpendicular to the string) of the two masses, M and m, fixed on a massless wire of length 4a. the entire system lies on a frictionless table.

Exercise 10.7.

A cylinder (of mass M_1. Radius R and height h) rests on a massless disc and rotates around a fixed axis at the center of the disc (disc radius -D). at the rin of the disc is attached a point mass M_2. There is friction between the cylinder and disc. Lat $D - 2R$ and M_1-$2M_2$. The dimensionless coefficient of kinetic friction is c, and the acceleration of gravity is g. the initial angular velocity of the cylinder (ω_1^0) is four times that of the disc (ω_2^0), i.e. ω_1^0- $4\omega_2^0$. In terms of R, M_1, σ and g only, find

(A) The time t required for the system to reach a steady state.
(B) The final angular velocity of the disc and the cylinder.

Exercise 10.8.

A string of length L is fixed at both ends, has total mass M, and is stretched under tension T. At time $t = 0$, the string is struck by a hammer of width d at position $x = a$ (see diagram) in such a way as to set the string vibrating with initial conditions.

$y(x, t = 0) = 0$ all x
$\dot{y}(x, 0) = 0$ $0 \le x \le a - \dfrac{d}{2}$
$\dot{y}(x, 0) = v_0$ $a - \dfrac{d}{2} \le x \le a + \dfrac{d}{2}$
$\dot{y}(x, 0) = 0$ $a + \dfrac{d}{2} \le x \le L$

(a) Find an expression for the (time-dependent) kinetic energy of the nth normal-mode of vibration of the string in the $\hat{y}$ direction. (There is no longitudinal vibration). Express the velocity and frequency of the wave in terms of the constants given in the problem.
(b) Find a position $x = a$ and width d of the hammer which will maximize the energy in the $n = 3$ mode of vibration.

Exercise 10.9.

A particle is constrained to move on the cycloid:
$$x = a\cos^{-1}\left(\frac{a-y}{a}\right) + \sqrt{2ay - y^2} \quad (0 \le y \le 2a)$$
Under the influence of gravity (the y-axis points upward).
(i) Write Lagrangian for this system.
(ii) Obtain the Euler equation(s).

(iii) Suppose the particle starts from a point $y = y_0$ with zero initial velocity: show that the time it takes to reach the bottom of the curve $(y = 0)$ Is independent of y_0.

$$\left[You\ may\ need\ the\ integral\ \int \frac{du}{\sqrt{u - u^2}} = sin^{-1}(2u - 1)u \right.$$
$$\left. < 1 \right]$$

Exercise 10.10.

(a) In the decay
$$A + p + \pi^-$$
What is the pion's energy, measured in the rest frame of the A? (*Find E_π in terms of the rest masses m_A, m_p, m_π*).

(b) A neutron with energy 939×10^{10} MeV travels across a galaxy whose diameter is 10^5 light years. If the half-life of a neutron is 640 s.. should you bet that the neutron will decay before it crosses the galaxy? (Justify your answer.)
$$m_n = 939\ MeV \qquad 1\ year = \pi \times 10^7\ 5.$$

Exercise 10.11.

The metric describing a spherical shell of matter of radius R can be written
$$ds^2 = -\left(1 - \frac{2M}{r}\right)dt^2 + \left(1 - \frac{2M}{r}\right)^{-1}dr^2$$
$$+r^2(d\theta^2 + sin^2\theta d\phi^2).\ outside$$
$$ds^2 = -dt^{-2} + dr^{-2} + r^{-2}(d\theta^2 + sin^2\theta d\phi^2).\ inside.$$

a) Find functions $\bar{t}(r, t), \bar{r}(r, t)$ near $r = R$, for which the metric is continuous at $r = R$.
b) A neutrino, emitted by a decaying neutron at the center of the shell $(\bar{r} = 0)$. Has energy E measured by an observer at rest at $\bar{r} = 0$. What is its energy when it reaches infinity $(r \gg R)$, measured by an observer at infinity? (It passes through the shell without interaction.)

Exercise 10.12.

A particle of mass m and charge e moves In a magnetic field $\underset{B}{\to}=$ $b(x^2 + y^2)\hat{k}$, where b is a constant.

(a) Find a vector potential for $\vec{B}$ of the form $\vec{A} = f(x^2 + y^2)\,\vec{\phi}$, where $\vec{\phi} = x\hat{j} - y\hat{i}$.

(b) Find the Hamiltonian for the particle, using this $\vec{A}$.

(c) Show that p_ϕ is a constant of the motion by verifying that the Poisson bracket $\left[p_\phi , H \right]_{PB}$ vanishes.

(d) Find a conserved quantity other that H and p_ϕ.

Exercise 10.13.

Consider the following three ways in which you could start out with a y-ray photon of energy 3 Mev and end up with a moving electron. Calculate the numerical value of the maximum kinetic energy that an electron could have in each case.

(a) Photoelectric effect

(b) Electron pair production

(c) Compton scattering (Derive any expression you use for Compton scattering.)

$$H = 6.63 \times 10^{-34} J \times s$$
$$= 4.136 \times 10^{-15} \ eV \times s$$

If you need more data that you do not know, make an estimate (of reasonable magnitude, if possible) and use that value for your calculation. Be explicit about the estimate you are using.

Exercise 10.14.

A relativistic collision takes place along a straight line between a particle of rest mass m_0 and another of rest mass nm_0. They stick together after the collision, and have a combined rest mass of M_0, which leaves with speed v. before the collision, m_0 is at rest and the other particle approaches at speed u. if we call

$$Y = \frac{1}{\sqrt{1 - \dfrac{u^2}{c^2}}}$$

Then find

A) V as a function of u and y. And

B) $\dfrac{M_0}{m_0}$ as a function of u and y.

Exercise 10.15.

173

In Eddington-Finkelstein coordinates the metric of a Schwarzschild black hole is

$$ds^2 = -\left(1 - \frac{2M}{r}\right) dv^2 + 2\, dv dr + r^2\{d\theta^2 + sin^2\theta d\phi^2\}.$$

(a) show that the M=0 case is flat space by finding a chart (coordinate system)

$t, r, \to \theta, \phi$ for which the metric (1) has the form

$$ds^2 = -dt^{-2} + dr^{-2} + r^{-2}(d\theta^2 + sin^2\theta d\phi^2)\ (M = 0).$$

(b) Let r(v) be a radial time like curve whose initial point lies within the horizon r(0) < 2M. show that r(v) < r(0) when v > 0 (i.e., the curve cannot emerge from the horizon).

(c) A flashlight and an observer, both on the $\theta = \phi = 0$ axis, are at fixed radii $r = r_f$ and $r = r_o$. The flashlight emits light of wavelength λ (measured In its frame). What wave-length does the observer measure?

(d) Show that the v = constant surfaces are null, $g^{ab\nabla} a^{v\nabla} b^v = 0$

Exercise 10.16.

A particle with charge 2 q moves in the electromagnetic field of a fixed particle that carries both an electric charge Q and a magnetic charge b: the fixed particle's magnetic field is

$$B = \frac{b\,\vec{r}}{r^3}$$

Prove that the vector

$$\vec{L} - \frac{qb}{c}\frac{\vec{r}}{r}$$

Is a constant of motion for the particle q, where $\underset{L}{\to}$ is the orbital angular momentum.

Exercise 10.17.

In the double pendulum shown, point masses 3m and m are connected by weightless rods of length l to each other and to a point of support. The masses are free to swing in a vertical plane.

At time $t = d, \theta = 0, \frac{d\theta}{dt} = 0, \phi = \phi_0 \ll 1\ and\ \frac{d\phi}{dt} = 0.$

Find $\theta(t)\ and\ \phi(t).$

Chapter 11. Series Outlook

The classic formulations of point particle motion have been described: using differential equations (Newton's 1st and 2nd Law); using a variational function formulation to select the differential equation (Lagrangian variation); using a variational functional formulation (Action formulation) to select the variational function formulation. Also described were the two domains for motion in many systems: non-chaotic; and chaotic.

From the Lagrangian variational formulation of 'action' for particle motion we will eventually define the path integral functional variational formulation involving that same Lagrangian to arrive at a quantum description for the non-relativistic quantum particle motion (described in detail in Book 4 [42], and relativistic in Book 5 [43]). From the quantum description we arrive at the propagator formalism for describing dynamics (this exists in the classical formulation too, but typically is not used much in that context). Complex propagators will then be found to have ties to statistical mechanics and thermodynamics properties (Book 6 [44]). The ties to statistical mechanics are further emphasized when at the "edge of chaos" but with the orbit motion still confined. This may be associated with an equilibrium and martingale regime, the existence of which can then be used at the start of Book 6 [44] statistical mechanics and thermodynamics derivations with the existence of equilibria established at the outset. The existence of the familiar entropy measures are already indicated in the neuromanifold description (Book 3 [41]), thus, together with equilibria, the Book 6 thermodynamics description is able to begin with a well-established foundation that is not claimed by fiat, rather claimed as a direct result of what has already been determined in the theory/experiment described in the previous Books of the Series [X].

When moving from a theory of point particles to a theory of fields, there's not much discussion in the core physics books on fields in a general sense, it usually just directly jumps to the main field of relevance, Electromagnetism (EM). If advanced, it may also cover General Relativity (GR), as with [92]. In the next two books in the Series [X] we will cover these topics, but we will also cover basic fields in 1, 2, and 3D (including fluid dynamics), as well as 4D Lorentzian Field formulations (for Special Relativity), the Gauge Field formulation (thus Yang Mills covered in a classical context), and the GR geometric and gauge formulations. This establishes the foundation for the standard forces, and upon quantization (Books 4 and 5 in the Series), lays the foundation for the standard renormalizable forces (all but gravitation).

In Book 2 the focus is on classical field theory in a fixed geometry, the main physical example is EM. In this setting alpha appears, for example, in the description of an electron-positron pair: $F = e^2/(4\pi\varepsilon a^2)$ for electron-positron distance 'a' apart, where alpha appears as the coupling constant. Later, in quantum mechanics, both modern and in the early Bohr model, we have that alpha = $[e^2/(4\pi\varepsilon)]/(c\hbar)$. The appearance of alpha in the situations is occurring in bound systems. If we examine EM interactions that are unbound, on the other hand, such as with the Lorentz Force $F = q(E \times v)$, here there arises no alpha parameter, nor with the early quantum mechanical analysis of such systems such as with Compton scattering. Thus, we see an early role for alpha, but only in bound systems, thus only in systems with (convergent) perturbative expansions in system variables.

In Book 3, classical field theory with *dynamic* geometry, i.e. GR, we don't see alpha at all. Instead we see manifold constructs and the mathematics of differential geometry (and to some extent differential topology and algebraic topology). Manifold constructs are described in the math background given in Book 3 and the Appendix. An application in the area of neuromanifolds (see [24]), shows the equivalent of a geodesic path in this setting is evolution involving minimum relative entropy steps. Similar to the description of a locally flat space-time we will find a description of 'entropy' increasing/evolving according to minimum relative entropy.

Appendix

A. A synopsis of Ordinary Differential Equations (ODE's)
This synopsis is at the level of Caltech's graduate course in applied mathematics AMa101 ca. 1985, where the main text used was by Bender & Orszag [39]. Many problems were assigned and complete solutions are provided for many of these problems. Thus, indirectly, solutions to several problems presented in [39] are included in what follows as well. The core material on differential equations and worked examples is selected to quickly educate on the amazing complexity possible and to clarify standard solution methods.

This synopsis includes an Introduction to ODEs; local ODE analysis (a study of singular points); nonlinear ODEs; Perturbation Methods (including WKB theory); and Sturm-Liouville Theory. The latter two topics are most relevant to problems in quantum mechanics (QM), so are placed as an appendix to Book 4 on QM.

A.1 Introduction to ODEs
Define an nth order ordinary differential equation (ODE) to be:

$$\frac{d^n y}{dx^n} = F\left(x, y, \frac{dy}{dx}, \dots, \frac{d^{n-1}y}{dx^{n-1}}\right) \rightarrow y^{(n)} = F\left(x, y, y^{(1)}, \dots, y^{(n-1)}\right),$$

$$(A-1)$$

and there is the alternate notation $y' = y^{(1)}; y'' = y^{(2)}$; etc., as well. If F is linear in $y, y^{(1)}, \dots, y^{(n-1)}$, then the ODE is a linear ODE [39]. The solution of an nth order linear ODE is a function of n constants of integration. If F is nonlinear there are still n constants of integration but there may be additional solutions which cannot be constructed by choosing the constants. Linear ODE's are often written in "operator notation":

$$\mathcal{L}\, y(x) = f(x),$$

$$(A-2)$$

where $\mathcal{L}$ is the differential operator:

$$\mathcal{L} = p_o(x) + p_1(x)\frac{d}{dx} + \cdots + p_{n-1}(x)\frac{d^{n-1}}{dx^{n-1}} + \frac{d^n}{dx^n}.$$

$$(A-3)$$

If $f(x) = 0$, then it is homogeneous, otherwise it is nonhomogeneous (having homogenous solutions plus particular solutions). We have an initial value problem (IVP) if we know $y, y^{(1)}, \ldots, y^{(n-1)}$ at some (initial) value $x = x_0$: $y(x_0) = a_0$, $y'(x_0) = a_1$, $\ldots$, $y^{(n-1)}(x_0) = a_{n-1}$, for which there is a general solution $y(x) = \sum_{j=1}^{n} c_j y_j(x)$, where the c_j are arbitrary constants of integration and the $\{y_j\}$ are a set of linearly independent solutions. To determine if our set of solutions are truly independent we must evaluate their Wronskian [39]. The Wronskian also arises naturally in addressing the IVP, so that will be considered next. Note, unlike IVP's, for a boundary value problem (BVP) we pose values (and/or derivatives) at more than one point. This is necessarily a global context of solution, not local, thus more complicated.

To show the existence and uniqueness for IVP's with $y^{(n)} = F(x, y, y^{(1)}, \ldots, y^{(n-1)})$ we can always convert the nth order equation to a system of n first order equations:

$$\frac{dy_i}{dx} = f_i(y_1, y_2, \ldots, y_n, x), \quad i = 1..n, \quad \text{where } y_i = \frac{d^{i-1}}{dx^{i-1}} y(x).$$

(A-4)

This is often written in vector notation:

$$\vec{Y} = \begin{pmatrix} y_1(x) \\ \ldots \\ y_n(x) \end{pmatrix}, \quad \vec{F} = \vec{F}(\vec{Y}, x) = \begin{pmatrix} f_1(x) \\ \ldots \\ f_n(x) \end{pmatrix}, \quad \frac{d\vec{Y}}{dx}$$

$$= \vec{F}(\vec{Y}, x), \quad \text{with IVP: } \vec{Y}(x = x_0) = \vec{Y_0}$$

(A-5)

To solve this we use a recursive approximation (Picard iteration) starting with the integral form:

$$\vec{Y}(x) = \vec{Y_0} + \int_0^x F(Y, t) dt.$$

(A-6)

Assuming $x_0 = 0$ without loss of generality (w.l.o.g.), we write:

$$\vec{Y_0}(x) = \vec{Y_0}; \quad \vec{Y_1}(x) = \vec{Y_0} = + \int_0^x \vec{F}(\vec{Y}, t) dt; \quad \ldots\ldots; \quad \vec{Y_{n+1}}(x)$$

$$= \vec{Y} + \int_0^x \vec{F}(\vec{Y_n}, t) dt.$$

(A-7)

Convergence of the sequence depends on $\vec{F}$. Let's show that the iteration converges in some neighbourhood of $x = 0$. First. let's show that $\vec{F}$ satisfies a Lipschitz condition:

178

$$\left\| \vec{F}(\vec{Y_1}, x) - \vec{F}(\vec{Y_2}, x) \right\| \le K \left\| \vec{Y_1} - \vec{Y_2} \right\|,$$

(A-8)

for all $\|\vec{Y} - \vec{Y_0}\| \le a$ and all $X: \|x\| \le b$. If working with pure numbers (or 1-dimension), have $\|x\| = |x|$, and, $|x - y| \ge 0$, with equality only when x=y. Also have $|x - y| = |y - x|$ (symmetry) and $|x - z| \le |x - y| + |y - z|$ (triangle inequality). For vectors: $\|\vec{x} - \vec{y}\| = |\sqrt{(\vec{x} - \vec{y}) \cdot (\vec{x} - \vec{y})}|$, and we still have symmetry and the triangle inequality. We also require that $\vec{F}$ be bounded:

$$\vec{F}(\vec{Y}, x) \le M.$$

If these conditions are satisfied then the Picard iteration converges. To demonstrate, consider:

$$\vec{Y_n}(x) = \vec{Y_0} + \int_0^x \vec{F}(\vec{Y}_{n-1}, t)dt \quad and \quad \vec{Y}_{n+1}(x) = \vec{Y_0} + \int_0^x \vec{F}(\vec{Y_n}, t)dt.$$

We then have:

$$\vec{Y}_{n+1} - \vec{Y_n} = \int_0^x \left[\vec{F}(\vec{Y_n}, t) - \vec{F}(\vec{Y}_{n-1}, t) \right]dt$$

$$\left\| \vec{Y}_{n+1} - \vec{Y_n} \right\| \le \int_0^x \left\| \vec{F}(\vec{Y_n}, t) - \vec{F}(\vec{Y}_{n-1}, t) \right\|dt \le K \int_0^x \left\| \vec{Y_n} - \vec{Y}_{n-1} \right\|dt.$$

To evaluate the RHS, consider:

$$\left\| \vec{Y_2} - \vec{Y_1} \right\| \le K \int_0^x \|Y_1 - Y_0\|dt \le K \int_0^x dt \int_0^t du \|F(Y_0, u)\|$$

$$\le KM \int_0^x dt \int_0^t du.$$

Using induction, it can be shown that:

$$\left\| \vec{Y}_{n+1} - \vec{Y_n} \right\| \le \frac{MK^n x^{n+1}}{(n+1)!}.$$

If we then write:

$$\vec{Y_n}(x) = \vec{Y_0} + \left(\vec{Y_1} - \vec{Y_2} \right) + \left(\vec{Y_2} - \vec{Y_3} \right) \cdots,$$

then, if the norm series converges, then $\vec{Y_n}$ will converge (it probably has negating factors):

$$\|\overrightarrow{Y_n}\| \leq \|\overrightarrow{Y_0}\| + \sum_{m=0}^{\infty} \frac{MK^m x^{m+1}}{(m+1)!} = \|\overrightarrow{Y_0}\| + \frac{M}{K}(e^{kx} - 1).$$

(A-9)

Thus we have a condition on the solution that is sufficient but not necessary. We need to show uniqueness to complete the general solution. We show uniqueness by counterexample, start with:

$$\vec{X} = \overrightarrow{X_0} + \int_0^x F(x,t)dt \quad and \quad \vec{Y} = \overrightarrow{Y_0} + \int_0^x F(y,t)dt ,$$

(A-10)

then

$$\|\vec{X} - \vec{Y}\| \leq \int_0^x \|F(\vec{X},t) - F(\vec{Y},t)\| \, dt \leq K \int_0^x \|\vec{X} - \vec{Y}\| dt$$

$$\leq K^2 \int_0^x dt \int_0^1 du \|\vec{X} - \vec{Y}\|,$$

thus

$$\|\vec{X} - \vec{Y}\| \leq \frac{K^{n+1}}{(n+1)!} \int_0^x (x - t)^n \|\vec{X} - \vec{Y}\| dt .$$

(A-11)

As n goes to infinity, the RHS goes to zero, and we see that $\|\vec{X} - \vec{Y}\| = 0$, and by the Lipschitz condition we then have $\vec{X} = \vec{Y}$, e.g., uniqueness. Thus, we see that a (unique) solution is generally possible. Practically speaking, what is this general solution?

General Homogeneous Solution (following notation of [39])
Consider:

$$\mathcal{L} \, y(x) = 0$$

(A-12)

As is usual with ODE's, let's consider a solution involving an exponential term: e^{rx}. substituting this as a trial function into the operator equation we get:

$$\mathcal{L} \, e^{rx} = e^{rx} P(r),$$

(A-13)

where $P(r)$ is an nth-order polynomial:

$$P(r) = r^n + \sum_{j=0}^{n-1} p_j r^j .$$

180

$$(A\text{-}14)$$

The solutions correspond to the zeros of $P(r)$, $r_1, r_2, \dots$, i.e. :

$$y = e^{r_1 x}, e^{r_2 x}, \dots$$

$$(A\text{-}15)$$

The only complication arises if there are repeated zeroes. Suppose the first root is m-fold, then we have a solution of the form:

$$\mathcal{L}\, e^{rx} = e^{rx}(r - r_1)^m\, Q(r),$$

$$(A\text{-}16)$$

where Q is a polynomial of degree $n - m$. A linear combination of all the solutions then constitutes a general solution.

General Inhomogeneous Solution

Consider the inhomogeneous equation,

$$\mathcal{L}\, y(x) = f(x).$$

$$(A\text{-}17)$$

One technique for finding a specific solution is known as variation of parameters, which works best if you have independent solution (non-zero Wronskian) (see [39]). Some examples involving this technique will e explored. In this quick synopsis we move on to considering Green's function methods to solving the inhomogenous equation. For this we make use of delta functions. For what follows we will define the delta function as:

$$\delta(x - a) = \begin{cases} 0 & x \neq a \\ \infty & x = a \end{cases},$$

$$(A\text{-}18)$$

such that:

$$\int_{-\infty}^{\infty} \delta(x - a)dx = 1 \quad and \quad \int_{-\infty}^{\infty} \delta(x - a)f(a)dx = f(x).$$

$$(A\text{-}19)$$

If we integrate part-way we get the classic Heaviside Step function (with step at x=a):

$$\int_{-\infty}^{\infty} \delta(x - a)dx = h(x - a).$$

$$(A\text{-}20)$$

The Green's function method is to then obtain the particular solution to

$$\mathcal{L}\, G(x, a) = \delta(x - a),$$

$$(A\text{-}21)$$

181

where the solution to the general inhomogeneous equation then trivially follows from:

$$y_p(x) = \int_{-\infty}^{\infty} da\, f(a) G(x, a).$$

(A-22)

In what follows, let's specialize to a second order differential equation (trivial 2x2 Wronskian). In which case we arrive at the form:

$$\frac{d^2}{dx^2} G(x, a) + p(x)\frac{d}{dx} G(x, a) + p_0(x) G = \delta(x - a).$$

(A-23)

Now, the L:HS must match the singularity of the delta function on the RHS. Thus, argue that $d^2 G/dx^2 \sim \delta(x - a)$ (thus G has to be less singular than $\delta(x - a)$. Likewise, we must have dG/dx no more singular than a step function, e.g., $dG/dx \sim h(x - a)$. Consistent with this is that G must be no more variant than a ramp function (zero until ramp starts at x=a), which will be denoted by 'r': $G \sim r(x - a)$. This is all we need to know to arrive at a general formulation of the solution. the trick is to now analyzr the ODE by integrating from $a - \varepsilon$ to $a + \varepsilon$ and let $\varepsilon \to 0$:

$$\int_{a-\varepsilon}^{a+\varepsilon} \frac{d^2 G}{dx^2} dx + \int_{a-\varepsilon}^{a+\varepsilon} p\frac{dG}{dx} dx + \int_{a-\varepsilon}^{a+\varepsilon} G p_0 dx = \int_{a-\varepsilon}^{a+\varepsilon} \delta(x - a) = 1.$$

Thus,

$$\left.\frac{dG}{dx}\right|_{a+\varepsilon} - \left.\frac{dG}{dx}\right|_{a-\varepsilon} = 1.$$

(A-24)

Working with two (independent) homogenous solutions, $y_1(x)$ and $y_2(x)$, we know we can express the inhomogeneous solution on either side of the singularity in the 'homogenous' form for that side. Let's write the Green's function in this way:

$$G(x, a) = \begin{cases} A_1 y_1(x) + A_2 y_2(x) & x < a \\ B_1 y_1(x) + B_2 y_2(x) & x \geq a \end{cases}$$

(A-25)

Since G is continuous at x=a we then have:

$$A_1 y_1(a) + A_2 y_2(a) = B_1 y_1(a) + B_2 y_2(a)$$
$$B_1 y_1'(a) + B_2 y_2'(a) - A_1 y_1{}'(a) - A_2 y_2{}'(a) = 1$$

In matrix notation:

$$\begin{bmatrix} y_1(a) & y_2(a) \\ y_1{}'(a) & y_2{}'(a) \end{bmatrix} \begin{bmatrix} B_1 - A_1 \\ B_2 - A_2 \end{bmatrix} = \begin{bmatrix} 0 \\ 1 \end{bmatrix},$$

which can be solved by

182

$$B_1 - A_1 = \frac{-y_2(a)}{W(y_1(a), y_2(a))}$$

$$B_2 - A_2 = \frac{y_1(a)}{W(y_1(a), y_2(a))}$$

where W is the Wronskian, which is

$$W = det \begin{bmatrix} y_1(a) & y_2(a) \\ y_1'(a) & y_2'(a) \end{bmatrix}.$$

Using this,

$$y(x) = \int_{-\infty}^{\infty} G(x, a) f(a) da$$

is the entire solution if $y(x)$ satisfies $Ly(x) = f(x)$ and $y(x)$ satisfies the BC's or initial values specified. Let's consider a simple example:

$$y'' = f(x) \quad with \quad \begin{matrix} y(0) = 0 \\ y'(1) = 0 \end{matrix}$$

We get $W = \begin{bmatrix} 1 & x \\ 0 & 1 \end{bmatrix} = 1$, and

$$B_1 - A_1 = -a$$
$$B_1 - A_1 = 1$$

Thus,

$$G(x, a) = \begin{cases} A_1 y_1(x) + A_2 y_2(x) & x < a \\ B_1 y_1(x) + B_2 y_2(x) & x \geq a \end{cases} = \begin{cases} A_1 + A_2 x & x < a \\ B_1 + B_2 x & x \geq a \end{cases},$$

(A-26)

from which we determine:

$$\begin{matrix} A_1 = 0 & B_1 = -a \\ B_2 = 0 & A_2 = -1 \end{matrix}.$$

Thus,

$$G = \begin{cases} -x & x < a \\ -a & x \geq a \end{cases}.$$

Solving for $y(x)$:

$$y(x) = \int_0^1 da\, G(x, a) f(a) = \int_0^a da\, (-x) f(a) + \int_a^1 da\, (-a) f(a)$$

(A-27)

Non-linear ODE's (see [65] for many examples)

For our first non-linear ODE, let's consider Bernoulli's equation:

$$y'(x) = a(x) y + b(x) y^p .$$

(A-28)

Let's try to solve by substituting $u(x) = y(x)^{1-p}$, where:

183

$$\frac{du}{dx} = (1-p)y^{-p}\frac{dy}{dx}.$$

(A-29)

We thus obtain:

$$\frac{du}{dx} = [a(x)y^{-p} + b(x)](1-p),$$

(A-30)

which is a first order ODE and thus directly solvable.

If we work with the same first order form, except now with quadratic in y, we get the Riccati equation. A simple transformation shows that the general Riccati equation relates to the general (linear) second order differential equation. Thus, we've already hit a limitation in obtaining general solutions even for the seemingly 'simple' Riccati equation. This is because a general solution to the linear second order differential equation does not exist (thus a general solution to the Riccati equation doesn't exist). That said, Let's try to solve the following Riccati equation:

$$y' = y^2 + \frac{y}{x} + x^2.$$

(A-31)

We find a solution with $y = x$, so let's consider a general solution of the form: $y = x + u(x)$:

$$u' = \left(2x + \frac{1}{x}\right)u + u^2$$

(A-32)

which is a first order equation, and thus solvable.

Some other techniques worth mentioning, starting with operator 'factoring'. Consider

$$\frac{d^2y}{dx^2} + p(x)\frac{dy}{dx} + q(x)y = f(x).$$

(A-33)

We can factor this as

$$\left(\frac{d}{dx} + a(x)\right)\left(\frac{dy}{dx} + b(x)\right)y = f(x).$$

(A-34)

The two forms are in agreement if $(b + a) = p$ and $b' + ab = q$.

Consider next the possibility of an 'exact' equation, e.g., where we have the form

$$M(x,y) + N(x,y)\frac{dy}{dx} = 0,$$

(A-35)

184

such that

$$M(x,y)dx + N(x,y)dy = dF(x,y) = \left[\frac{\partial F}{\partial x}\right]dx + \left[\frac{\partial F}{\partial y}\right]dy = 0.$$

Thus, the test for having an exact form is that

$$\frac{\partial M}{\partial y} = \frac{\partial N}{\partial x}.$$

(A-36)

Consider next the notion of 'integrating factor'. This situation arises if

$$M(x,y)dx + N(x,y)dy \neq dF(x,y),$$

but by multiplying thru by an (integrating) factor we find that:

$$\mu(x,y)M(x,y)dx + \mu(x,y)N(x,y)dy = dF(x,y).$$

He latter expression is then an exact form if

$$\frac{\partial(M\mu)}{\partial y} = \frac{\partial(N\mu)}{\partial x}.$$

(A-37)

For higher-order nonlinear ODE's important simplification are possible if specific forms exist, let's consider some of those:

(i) Autonomous – an ODE is autonomous if it does not have an explicit dependence on the dependent variable.

(ii) Equidimensional – an ODE is equidimensional if the substitution $x \rightarrow ax$ leaves the equation invariant. Such an equation can be trivially shifted to autonomous form with the substitution $x = e^t$.

(iii) Scale invariant – an ODE is scale invariant if the substitutions $x \rightarrow ax$ and $y \rightarrow a^p y$ leaves the equation. Such an equation can be trivially shifted to equidimensional form (and from there to autonomous) with the substitution $y = x^p u$. Let's now turn to the issue of singular points in solving ODE's.

The above methods of solution for ODE's are so robust that even when exact solutions can't be obtained approximate solutions can generally be obtained locally near a point of interest. Often this is all that is needed anyway. So the only thing that can go wrong is if the reference point of interest isn't 'ordinary', i.e., if the point is 'singular.' Let's now explore this possibility.

Singular points of homogeneous linear equations
Recall the notation introduced for the homogeneous linear differential equation:

$$\mathcal{L} y(x) = f(x),$$

185

where
$$\mathcal{L} = p_0(x) + p_1(x)\frac{d}{dx} + \cdots + p_{n-1}(x)\frac{d^{n-1}}{dx^{n-1}} + \frac{d^n}{dx^n}.$$

(A-38)

The general theory for the analysis of singular points begins with the above form when considering complex arguments, not just real [39,65, 66]. The theoretical results obtained [67] then categorize the singular points in terms of the analyticity (complex properties) of the coefficient functions:

Ordinary Point
A point x_0 is ordinary if all of the coefficient functions are analytic in the neighborhood of x_0. Fuchs showed in 1866 that all n linearly independent solutions for a nth order linear ODE (obtained from previous analysis methods) will be analytic in the neighborhood of an ordinary point.

Regular Singular Point
A point x_0 is a regular singular point if not all of the coefficient functions are analytic but if all of the terms in $\mathcal{L}\, y(x)$ are locally analytic (about the reference point x_0), i.e., when the following functions are analytic: $(x - x_0)^n p_0(x)$, $(x - x_0)^{n-1} p_1(x)$, ... , $(x - x_0)p_{n-1}(x)$. Note that a solution may be analytic at x_0 even if x_0 is a regular singular point. If not analytic at a regular singular point, a solution must either involve a pole or an algebraic or logarithmic branch point. Accordingly, Fuchs showed that there is always one solution of the form (following notation of [39]:
$$y = (x - x_0)^\alpha A(x),$$

(A-39)

where α is known as the indicial exponent and $A(x)$ is a function analytic at the regular singular point x_0. If the order is second or greater, then a second solution exists in one of two possible forms:
$$y = (x - x_0)^\beta B(x),$$

(A-40)

or
$$y = (x - x_0)^\beta B(x) + (x - x_0)^\alpha A(x) \ln(x - x_0).$$

(A-41)

Going to higher than second order, additional solutions have singular behavior, at its worst, of the form:
$$y = (x - x_0)^\delta \sum_{i=0}^{n-1} [\ln(x - x_0)]^i A_i(x),$$

(A-42)

186

where all the functions A_i are analytic. Thus, regular singular points can be handled in a comprehensive theory much like ordinary points.

Irregular Singular Point

A point x_0 is an irregular singular point if it isn't regular or ordinary. There is no comprehensive theory to use to solve if an irregular singular point. From Fuchs we know that if a complete set of solutions all had the forms indicated in the previous section, then the point must be regular. conversely, if we have an irregular singular point, then at least one of te solutions will not have the forms indicated above. Typically, in fact, the solutions all have essential singularities (not analytic) at the reference point x_0 where the irregular singular point (ISP) exists.

Example A.1.
$$x^2 y'' - x(x+1)y' + y = 0$$
we see that $x_0 = 0$ is irregular, try:
$$y(x) = \sum_{n=0}^{\infty} \frac{a_n}{x^{n+\alpha}}.$$

Then have:
$$y'(x) = -\sum_{n=0}^{\infty} (n+\alpha) \frac{a_n}{x^{n+\alpha+1}} \quad and \quad y''(x)$$
$$= \sum_{n=0}^{\infty} (n+\alpha)(n+\alpha+1) \frac{a_n}{x^{n+\alpha+2}}.$$

Thus
$$a_{n+1} = -(n+1)a_n \quad \rightarrow \quad y(x) = a_0 \sum_{n=0}^{\infty} \frac{(-1)^n n!}{x^n}.$$

So far our one solution isn't even good (it diverges) indicating some of the issues that can arise with irregular singular points (ISPs). The solution hints at an answer, however. Consider
$$y(x) = x \int_0^{\infty} \frac{e^{-t}}{x+t} dt .$$

Then we have:

$$x^2 y'' - x(x+1)y' + y$$
$$= \int_0^\infty e^{-t} \left[\frac{-2x^2}{(x+t)^2} + \frac{2x^2}{(x+1)^3} - \frac{x^2+x}{x+t} + \frac{x^3+x^2}{(x+t)^2} \right.$$
$$\left. + \frac{x}{x+t} \right] dt = 0,$$

which works. Working with the indicated solution, let's expand for $x \to \infty$:

$$y(x) = \int_0^\infty \frac{e^{-t}}{1+t/x} dt$$

let $t = xS$ to get:

$$y(x) = \int_0^\infty \frac{e^{-xs}}{1+S} ds \approx \sum_{n=0}^\infty \frac{(-1)^n n!}{x^n}.$$

Let's now consider the exponential behavior near the ISP for the following:

$$y'' - (x^2+1)y = 0$$

where the ISP is at $x_0 = \infty$. We have for solutions

$$y_1(x) = e^{x^2/2} \quad and \quad y_2(x) = e^{x^2/2} erfc(x) \approx \frac{1}{\sqrt{\pi}} \frac{1}{x} e^{\frac{x^2}{2}} \ as \ x \to \infty.$$

If $x_0 \neq \infty$ then typical behavior might be $\exp\left(-\frac{1}{(x-x_0)^2}\right)$. To determine leading behavior write:

$$y(x) = e^{S(x)}, \quad y' = S'e^{S(x)}, \quad and \quad y'' = [(S')^2 + S'']e^S.$$

Thus

$$S'' + (S') - (x^2+1) = 0 \ as \ x \to \infty.$$

Using the method of *__Dominant Balance__* :

Note that x^2 getting big, what is balancing it?
- (i) S'' gets big faster than $(S')^2$, and $S'' \gg (S')^2$ as $x \to \infty$.
- (ii) $S'' \ll (S')^2$ as $x \to \infty$ (always true at ISP).
- (iii)   All three terms are same order (bad, can't use method).

Consider case (i): $S'' \approx x^2$ as $x \to \infty$, which gives $S' \approx x^3/3$, but this is inconsistent with $S'' \gg (S')^2$ as $x \to \infty$.

Consider case (ii): $(S')^2 \approx x^2$ as $x \to \infty$, which gives $S' \approx \pm x$, thus $S'' \approx \pm 1$. Since $S'' \ll (S')^2$ as

$x \to \infty$ this is consistent. We see that $S \approx \pm x^2/2$ works. In fact,$+ x^2/2$ is an exact solution. For the other solution, let's try: $S(x) = -x^2/2 + C(x)$. This spawns a separate dominant balance analysis, and we find that the only valid choice is $C(x) \sim -\ln(x)$, and

188

$$S \sim -x^2/2 - \ln(x) + \cdots$$

Thus,

$$y(x) \sim e^{-\frac{1}{2}x^2} \sum_{n=1}^{\infty} a_n x^{-n} = e^{-\frac{1}{2}x^2} F(x)$$

and we can proceed with the classic Frobenius method from here [65]:

$$y'' - (x^2 + 1)y = e^{-\frac{1}{2}x^2}[F'' - 2xF' - 2F] = 0$$

Use standard series expansion for F:

$$0 \cdot a_1 + 2 \cdot a_2 + \sum_{n=3}^{\infty} [(n-2)(n-1)a_{n-2} + 2(n-1)a_n]x^{-n} = 0$$

Thus, we have that: a_1 is arbitrary, $a_2 = 0$, and $a_{n+2} = -\frac{n}{2}a_n$. Thus,

$$a_{2n+1} = \frac{(-1)^n(2n-1)!!}{2^n} a_1$$

$$y(x) \sim e^{-\frac{1}{2}x^2} \sum_{n=0}^{\infty} \frac{(-1)^n(2n-1)!!}{2^n x^{2n+1}} a_1.$$

Let's consider the systematic expansion means a regular singular point, specialized to second order:

$$\mathcal{L}y = y'' + \frac{p(x)}{x}y' + \frac{q(x)}{x^2}y = 0$$

Assume a regular singular point at x=0 and that p(x), q(x) are analytic about x=0. Substitute

$$y = \sum_{n=0}^{\infty} a_n x^{n+\alpha}.$$

Example A.2.
Solve:

$$y'' + \frac{1}{xy'} - \left(1 + \frac{v^2}{x^2}\right)y = 0.$$

We have: $p(x) = 1$, $p_0 = 1$, $q(x) = -x^2 - v^2$, $q_0 = -v^2$.
Thus,

At order $x^{\alpha-2}$; $(\alpha(\alpha-1) + \alpha - v^2)a_0 = 0 \rightarrow \alpha^2 - v^2 = 0 \rightarrow \alpha = \pm v$. If v is a fractional number ($v \neq 0$ and $2v \neq n$) we get two solutions, so done, and have:
At order $x^{\alpha-1}$: $x^{\alpha-1}[(\alpha+1)^2 - v^2]a_1 = 0 \rightarrow a_1 = 0$

189

At order $x^{\alpha+n-2}$: $x^{\alpha+n-2}[(\alpha+n)^2 - v^2]a_n = a_{n-2} \rightarrow 0 = a_1 = a_3 = a_5 \ldots$
The solution is thus:

$$y(x) = a_0 \Gamma(v+1)x^v \sum_{n=0}^{\infty} \frac{(x/2)^{2n}}{n!\,\Gamma(n+v+1)}.$$

Notice that $a_n = (a_n - 2)/[(-v+n)^2 - v^2]$. So, for $\alpha = -v$ the denominator vanishes when $n = 2v$. If v is half-integral i.e. $1/2, 3/2, \ldots$, then $2v$ is odd-integer. After $2v$ steps we have a new arbitrary constant a_{2v} (happens for Bessel functions for example) and the recursion relation then generates two linearly independent solutions.

Double root case: $\alpha_1 = \alpha_2$
Consider the Frobenius form for the first solution: $x^\alpha \sum_{n=0}^{\infty} a_n(\alpha)x^n = y(x,\alpha)$. When there is a double root it can be shown that a second solution follows from the relation (derived in [39]):

$$\mathcal{L}\left[\frac{\partial}{\partial\alpha}y(x,\alpha)\Big|_{\alpha=\alpha_1}\right] = 0.$$

Example A.3. The Modified Bessel Function for $v = 0$:

$$y'' + \frac{1}{x}y' - y = 0,$$

where there is a double root at $\alpha = 0$ upon substitution with the Frobenius form above. Evaluating at various orders:
We start with a_0 being an arbitrary constant.
At $\mathcal{O}(x^{\alpha-1})$ we have $[(\alpha+1)^2 a_1] = 0 \rightarrow a_1 = 0$.
At $\mathcal{O}(x^{\alpha+n-2})$ we have $[(\alpha+n)^2 a_n - a_{n-2}] = 0$, thus, for $n \geq 2$ we have

$a_2 = \dfrac{a_0}{(\alpha+2)^2}$

$a_4 = \dfrac{a_0}{(\alpha+4)^2(\alpha+2)^2}$

$a_4 = \dfrac{a_0}{(\alpha+6)^2(\alpha+4)^2(\alpha+2)^2}$

Thus, we have for one solution (for $\alpha = 0$):

$$I_0(x) = a_0\left[1 + \frac{(x/2)^2}{(1!)^2} + \frac{(x/2)^4}{(2!)^2}\cdots\right] = a_0 \sum_{n=0}^{\infty}\frac{(x/2)^{2n}}{(n!)^2}.$$

The other solution is $\dfrac{\partial}{\partial\alpha}x^\alpha \sum_{n=0}^{\infty} a_n(\alpha)x^n\Big|_{\alpha=0}$. The other solution is then:

190

$$y(x) = \ln x\, I_0(x) + \sum_{n=0}^{\infty} \frac{\partial}{\partial \alpha} a_n(\alpha)\Big|_{\alpha=0} x^n = \ln x\, I_0(x) + \sum_{n=0}^{\infty} b_n x^n$$

$$= K_0(x).$$

In general, we see that the odd b_n vanish (like with a_n), and for even n:

$$b_{2n} = \frac{-a_0}{2^{2n} n!}\left[1 + \frac{1}{2} + \frac{1}{3} + \frac{1}{4} + \cdots \frac{1}{n}\right].$$

For further discussion of the modified Bessel solutions, for v = integer, see [39] and the worked examples that follow.

Using Dominant Balance to solve inhomogeneous equations
Example A.4.

$$y' + xy = 1/x^4$$

Consider the asymptotic behavior as $x \to 0$:

(1) Balance $y' + xy \sim 0$ *asymptotic to zero(authors don'tlike)*
   This has y asymptotic to zero, which is inconsistent with
   $y \sim A\exp(-x^2/2) \to 0$.
(2) $xy \sim 1/x^4 \to y \sim 1/x^5$ (which is inconsistent).
(3) $y' \sim \frac{1}{x^4} \to y = -\frac{1}{3}x^{-3}$, which is consistent with $xy \sim x^{-2}$.

So, try: $y = -\frac{1}{3}x^{-3} + C(x)$, which is balanced if $C = -\frac{1}{3}x^{-1}$ for the solution.

Example A.5. (Inhomogeneous Airy Equation)

$$y'' = xy - 1$$

where we consider the asymptotics for $y(x \to +\infty) \to 0$. This can be solved by variation of parameters. Since second order, have two independent solution types for homogeneous Airy equation, let's denote them by:

$$y_1 = Ai(x), \qquad y_2 = Bi(x).$$

The general solution by variation of parameters is thus

$$y(x) = \pi\left[Ai(x) \int_0^x Bi(t)dt + Bi(x) \int_x^{\infty} Ai(t)dt\right] + CAi(x)$$

The Asymptotic behavior of Ai, Bi is:

$$Ai(x) \sim \frac{1}{2\sqrt{\pi}} x^{-1/4} \exp\left(-\frac{2}{3}x^{\frac{3}{2}}\right)$$

$$Bi(x) \sim \frac{1}{\sqrt{\pi}} x^{-1/4} \exp\left(-\frac{2}{3}x^{\frac{3}{2}}\right)$$

Thus,

191

$$\int_0^x Bi(t)dt \sim \int_0^x \frac{1}{\sqrt{\pi}} t^{-1/4} \exp\left(\frac{2}{3}t^{3/2}\right) dt$$

$$= \int_0^x \frac{1}{\sqrt{\pi}} t^{-\frac{1}{4}} t^{-\frac{1}{2}} \frac{d}{dt} \exp\left(\frac{2}{3}t^{3/2}\right) dt$$

$$\int_0^x Bi(t)dt \sim \frac{1}{\sqrt{\pi}} x^{-3/4} \exp\left(2/3 \, x^{3/2}\right) + \cdots$$

$$\int_x^\infty Ai(t)dt \sim \int_x^\infty \frac{1}{2\sqrt{\pi}} t^{-1/4} \exp\left(-\frac{2}{3}t^{3/2}\right) dt$$

$$= \frac{1}{2\sqrt{\pi}} x^{-3/4} \exp\left(-2/3 \, x^{3/2}\right) + \cdots$$

Thus,

$$y(x) = \pi \frac{1}{2\sqrt{\pi}} x^{-1/4} \exp\left(-\frac{2}{3}x^{3/2}\right) \frac{1}{\sqrt{\pi}} x^{-3/4} \exp\left(\frac{2}{3}x^{3/2}\right) + $$
$$\pi \frac{1}{\sqrt{\pi}} x^{-1/4} \exp\left(\frac{2}{3}x^{3/2}\right) \frac{1}{2\sqrt{\pi}} x^{-3/4} \exp\left(-\frac{2}{3}x^{3/2}\right)$$
$$+ C \, Ai(x)$$

which simplifies to simply be:

$$y(x) \sim \frac{1}{x}.$$

Let's repeat the analysis using dominant balance method:
Consider $y'' \sim -1 \to y \sim -x^2/2$, which is inconsistent.
Consider $-xy \sim -1 \to y \sim \frac{1}{x}$, which is consistent, and done.

So far we've obtained the first order behavior, let's now consider the correction term:
$y = 1/x + C(x) \to y = -1/x^2 + C' \to y'' = 2/x^3 + C''$, so upon substitution we have:

$$\frac{2}{x^3} + C'' - 1 - xC(x) = -1 \to C'' - xC \sim -\frac{2}{x^3}$$

192

A separate dominant balance on the last expression reveals consistency with $C(x) \sim \frac{2}{x^4}$. We thus have the first two orders, lets write the general solution in the form:

$$y(x) \sim \frac{1}{x} \sum_{n=0}^{\infty} a_n x^{-3n} \qquad as \ x \to \infty$$

Suppose

$$y(x) = \frac{1}{x} \sum_{n=0}^{\infty} a_n x^{-3n}$$

then

$$y'(x) = -\frac{1}{x^2} \sum a_n x^{-3n} + \frac{1}{x} \sum (-3n) a_n x^{-3n-1}$$

$$y''(x) = \frac{2}{x^3} \sum a_n x^{-3n} - \frac{2}{x^2} \sum_{n=0}^{\infty} a_n (-3n) x^{-3n-1} + \frac{1}{x} \sum (-3n) a_n x^{-3n-2}$$

Thus, from $y'' - xy = -1$ we have:

$$\sum_{n=0}^{\infty} (2 + 6n + (3n)(3n+1)) a_n x^{-3n-3} - \sum_{n=0}^{\infty} a_n x^{-3n} = -1$$

The coefficient relations are then:

$$a_0 = 1$$

and

$$a_{n+1} = (3n+1)(3n+2) a_n$$

Thus,

$$y(x) = \frac{1}{x} \sum_{n=0}^{\infty} \frac{(3n)!}{3^n (n!)} \frac{1}{x^{3n}}$$

Example A.6.

Let's now consider an example where balancing only 2 terms fails:

$$y' - \frac{y}{x} = \frac{\cos x}{x^2} \qquad want \ behaviour \ as \ x \to 0^+$$

Try to balance with $y' - y/x \sim 0 \to y' \sim cx$ (inconsistent).

Try to balance with $-\frac{y}{x} \sim \frac{\cos x}{x^2} \to y \sim \frac{-\cos x}{x}$ (inconsistent).

Try to balance with $y' \sim \frac{\cos x}{x^2} \to y \sim -\frac{1}{x}$ (also inconsistent, but close)

So we move to a three term dominant balance with $\cos x \to 1$:

$$y' - \frac{y}{x} \sim \frac{1}{x^2} \to y \sim \frac{C}{x} \to y \sim -\frac{C}{x^2}$$

which is consistent for $C = -1/2$.

193

Nonlinear differential equations have pole positions dependent on initial conditions (cannot be found by inspection). In general, even if the equation is both regular and the Picard theorem guarantees a solution locally, it is still hard to know where with nearest singularity is. For example, consider:

$$y^1 = \frac{y^2}{1 - xy} \qquad y(0) = 1$$

Substitute with $y = \sum_{n=0}^{\infty} a_n x^n \rightarrow a_n = \frac{(n+1)^{n-1}}{n!}$. WE can now evaluate the radius of convergence R:

$$R = \lim_{n \to \infty} \left| \frac{a_n}{a_{n+1}} \right| = \lim_{n \to \infty} \left| \frac{n+1}{n+2} \frac{(n+1)^{n-2}}{(n+2)^{n-1}} \right| = \lim_{n \to \infty} \left| \left(1 - \frac{1}{n+2}\right)^n \right| = \frac{1}{e}.$$

Let's now consider a second-order differential equation having 'Sturm-Liouville' (S-L) form:

$$\frac{d}{dz} p \frac{d\Psi}{dz} + (q + \lambda R)\Psi = 0 \quad with \quad BC's \quad \Psi(a) = \Psi(b)$$
$$= 0 \qquad a < z < b.$$

$$(A\text{-}43)$$

Properties of the S-L equation:
- No solutions in general unless $\lambda = \lambda_m$, $\Psi = \Psi_m$
- The λ_m are rounded from below and it is always possible to adjust things so that $\lambda_0 = 0$
- The $\lambda_m's \to +\infty$ as $n \to \infty$
- $\int_a^b R(z)\Psi_n(z)\Psi_m(z)dz = E_n^2 \delta_{nm}$
- Claim: We can use the eigenfunctions to fit an arbitrary function in a least squares sense:

$$f(z) = \sum_{n=0}^{\infty} A_n \Psi_n(z),$$

$$(A\text{-}44)$$

where

$$\int_a^b R(z)f(z)\Psi_m(z)dz = \sum_{n=0}^{\infty} A_n \int_a^b dz\, R\, \Psi_n \Psi_m = A_n E_n^2.$$

$$(A\text{-}45)$$

Thus,

$$A_n = \frac{\int_a^b R(z)f(z)\Psi_m(z)dz}{E_n^2}.$$

194

Thus, we are claiming that $\sum_{n=0}^{N} A_n \Psi_n(z)$ is a solution to the problem of finding a lead squares fit to $f(z)$. To prove this we would like to minimize $I = \int_a^b R(z)dz[f(z) - \sum_{n=0}^{N} A_n \Psi_n(z)]^2$:

$$\frac{\partial I}{\partial A_m} = 0 = \int_a^b R(z)dz \left[f(z) - \sum_{n=0}^{N} A_n \Psi_n(z) \right] \left[-\sum_{n=0}^{N} \delta_{nm} \Psi_n(z) \right].$$

We want to show that as $N \to \infty$ the error, in a least squares sense, goes to zero. We can show that solving a Sturm-Liouville is equivalent to minimizing:

$$\Omega = \int_a^b \left[p(z) \left(\frac{d\Psi}{dz} \right)^2 - q(z) \Psi^2 \right] dz$$

Subject to $\int_a^b \Psi^2 R(z)dz = constant$. Suppose we choose a trial function $\Psi(z)$ which satisfies the B.C.'s at $z = a, b$ and normalized so that

$$\int_a^b R(z)dz \, \Psi^2(z) = 1$$

Compute:

$$\Omega(\Psi_0) = \int_a^b \left[p \left(\frac{d\Psi_0}{dz} \right)^2 - q \Psi_0^2 \right] dz$$

$$= \left[p \Psi_0 \frac{d\Psi_0}{dz} \right]_a^b - \int_a^b \Psi_0 \left[\frac{d}{dz} \left(p \frac{d\Psi_0}{dz} + q \Psi_0^2 \right) \right]$$

Thus

$$\Omega(\Psi_0) = \int_a^b \Psi_0 R \lambda_0 \Psi_0 dz = \lambda_0$$

(where λ_0 is typically the lowest eigenvalue). Similarly, with $\Psi = \sum_{n=0}^{N} A_n \Psi_n(z)$ we get:

$$\Omega(\Psi) = \int_a^b Rdz \sum_{n=0}^{N} A_n \Psi_n \sum_{m=0}^{M} \lambda_m A_m \Psi_m = \sum_{n=0}^{N} A_n^2 \lambda_m E_N^2 .$$

To complete the proof using the above we need to show that the least squares error decreases with N, but that is left to the references [65].

Asymptomatic appropriations for S-L eigenfunctions and eigenvalues
Recall the S-L equation:

$$\frac{d}{dz} p \frac{d\Psi}{dz} + (q + \lambda R)\Psi = 0$$

(A-49)

Let's make an 'inspired transformation':

$$y = (pR)^{1/4}\Psi$$

(A-50)

and define new values:

$$\varepsilon = \frac{1}{J}\int_a^z \sqrt{\frac{R}{P}}\, dz \quad and \quad J = \frac{1}{\pi}\int_a^b \sqrt{\frac{R}{P}}\, dz \, .$$

(A-51)

The S-L equation then becomes solvable in terms of the Volterra Integral equation:

$$\frac{d^2y}{d\varepsilon^2} + \left(k^2 + \omega(\varepsilon)\right)y(\varepsilon) = 0,$$

(A-52)

where

$$k^2 = J^2\lambda \quad and \quad \omega = \left[\frac{1}{(pR)^{1/4}}\frac{d^2}{d\varepsilon^2}(pR)^{1/4} - J^2\frac{q}{R}\right],$$

(A-53)

and we have $a < z < b$ (as before) and $0 < \varepsilon < \pi$. Solutions can be written:

$$y(\varepsilon) = A\sin(k\varepsilon) + B\cos(k\varepsilon) + \frac{1}{k}\int_{\varepsilon_0}^{\varepsilon} \sin(k(\varepsilon - t))\, w(t)y(t)dt.$$

Suppose $\Psi(a) = \Psi(b) = 0$, then $k = n$ and

$$\Psi_n \sim \frac{1}{(Rp)^{1/4}}\sin(n\varepsilon) \quad and \quad \lambda_n = \left(\frac{n}{J}\right)^2$$

Suppose we have general B.C.'s $\alpha\Psi + \beta\frac{d\Psi}{dz} = 0$ at $z = a, b$, then we have

$$k_n \sim \frac{J}{\pi n}\left[\frac{\alpha}{\beta}\sqrt{\frac{P}{R}}\right]_a^b$$

196

Example: the Singular S-L with $p(a) = 0$ *or* $p(b) = 0$ *or both* such as occurs with the Bessel equation:

$$\frac{d}{dz}\left(z\frac{d\Psi}{dz}\right) + \left(\lambda z - \frac{m^2}{z}\right)\Psi = 0,$$

(e.g., the S-L equation with $p = z$; $R = z$; and $q = -m^2/z$). Here, the singular point is $z = 0$ and we have:

$$\Psi = \frac{1}{\sqrt{z}}y, \quad J = \frac{1}{\pi}\int_0^b dz = \frac{b}{\pi}, \quad \varepsilon = \frac{\pi z}{b}, \quad k^2 = \frac{b^2\lambda}{\pi^2}$$

to give:

$$\frac{d^2y}{d\varepsilon^2} + \left[k^2 - \frac{(m^2 - 1/4)}{\varepsilon^2}\right]y = 0$$

with solutions:

$$y(\varepsilon) = \cos(k\varepsilon + \theta) - \frac{1}{k}\int_\varepsilon^\infty \sin(k(\varepsilon - t)y(t)\left(\frac{m^2 - 1/4}{t^2}\right)dt$$

Bessel functions have local behavior of the form
$z^{\pm m}[Taylor\ series\ in\ z]\ \ and\ \ J_n \sim z^n[\sum A_n z^{2n}]$.

A.2 ODE's having Sturm-Liouville form – asymptotic approximations
(Some of this material was covered in Ama101b in Spring of 1986.)

Example A.7. Verify Abel's formula for the Wronskian. That is, show that if

$$\frac{d^n y}{dx^n} + p_{n-1}(x)\frac{d^{(n-1)}y}{dx^{(n-1)}} + \cdots p_0(x)y(x) = 0$$

then the Wronskian W(x) satisfies

$$\frac{dW}{dx} = -p_{n-1}(x)W(x).$$

Solution
When we take the derivative of the Wronskian, we distribute to get derivatives inside the determinant on a row-by-row basis. This make two rows the same on all but the determinant with its derivative in the last row. If we then consider $\frac{dW}{dx} + p_{n-1}(x)W(x)$ we see both terms

contributing polynomial expressions involving y_n^n and $p_{n-1}y_n^{n-1}$, such that regrouping in a new determinant is possible with these terms grouped in the new last row, such as $y_n^n + p_{n-1}y_n^{n-1}$ is the last element of the last row, for example. Since $(y_n^n + p_{n-1}y_n^{n-1}) + \cdots + p_0 y_0 = 0$, there is a clear dependence on the grouping in terms of lower-order elements (obtainable from grouping of other rows), thus this determinant will be zero, and we have:

$$\frac{dW}{dx} + p_{n-1}(x)W(x) = 0$$

as desired.

Example A.8. Find the formula for the Green's function of a third order in homogeneous linear equation. Generalize this formula to n^{th} order.

Solution
There are three conditions:
(i) G is continuous at $x = a$.
(ii) dG is continuous at $x = a$.
(iii) $d^2 G|_{a^+} - d^2 G|_{a^-} = 1$
Thus,

$$\begin{bmatrix} y_1(a) & y_2(a) & y_3(a) \\ y_1'(a) & y_2'(a) & y_3'(a) \\ y_1''(a) & y_2''(a) & y_3''(a) \end{bmatrix} \begin{bmatrix} B_1 - A_1 \\ B_2 - A_2 \\ B_3 - A_3 \end{bmatrix} = \begin{bmatrix} 0 \\ 0 \\ 1 \end{bmatrix}$$

Cramers rule:

$$B_1 - A_1 = \frac{y_2(a)y_3'(a) - y_3(a)y_2'(a)}{\det W[y_1(a), y_2(a), y_3(a)]}, \quad etc.$$

Three more conditions can be chosen to specify the boundary conditions. For n^{th} order let W_j be W with the jth column replaced by a column vector with all zeroes except for the last row:

$$B_j - A_j = \frac{W_j}{\det W}$$

Example A.9. Find a closed form solution to the following Riccati equation:

$$xy' - 2y + ay^2 = bx^4.$$

Solution

Guess $y = \sqrt{b/a}\,x^2$ (indicated by dominant balance on last few terms), then test that it works, which it does. Thus, we have a Bernoulli equation by making the substitution

$$y(x) = \sqrt{\frac{b}{a}}x^2 + u(x).$$

Solving the standard Bernoulli equation, there is then the general solution:

$$y(x) = x^2\left(\sqrt{\frac{b}{a}} + \frac{2}{Ce^{\sqrt{ab}\,x^2} - \sqrt{\frac{a}{b}}}\right).$$

Example A.10. Legendre polynomials $P_n(z)$ satisfy the difference equation

$$(n+1)P_{n+1}(z) - (2n+1)z\,P_n(z) + n\,P_{n-1}(z) = 0$$

With $P_0(z) = 1$, $P_1(z) = z$.

a) Define the generating function $f(x,y)$ by

$$f(x,z) = \sum_{n=0}^{\infty} P_n(z)\,x^n$$

Show that $f(x,z) = (1 - 2xz + x^2)^{-1/2}$.

b) If $g(x,z) = \sum_{n=0}^{\infty}\frac{P_n(z)x^n}{n!}$ show that $g(x,z) = e^{xz}J_0\left(x\sqrt{1-z^2}\right)$ where J_0 is a Bessel function which satisfies: $ty'' + y' + ty = 0$ with $y(0) = 1$ and $y'(0) = 0$.

Solution

(a) $f(x,z) = \sum_{n=0}^{\infty} P_n(z)\,x^n = \sum_{n=0}^{\infty} P_{n+1}(z)\,x^{n+1} + P_0(z)$ (where $P_0(z) = 1$), while
$f'(x,z) = \sum_{n=0}^{\infty}(n+1)P_{n+1}(z)\,x^n$ and $f''(x,z) = \sum_{n=0}^{\infty}(n+1)(n+2)P_{n+2}(z)\,x^n$. Thus, if we shift the indexing the difference equation ($n \rightarrow n+1$), and multiply the recursion equation above by $(n+1)x^n$ with summation n=0 to ∞:

$$\sum_{n=0}^{\infty}[(n+1)(n+2)P_{n+2}(z)x^n - z(n+1)(2n+3)P_{n+1}(z)x^n$$
$$+ (n+1)^2 P_n(z)x^n] = 0$$

becomes:

$$f''(x,z) + \sum_{n=0}^{\infty} [-z[3(n+1) + 2n(n+1)]P_{n+1}(z)x^n + [n(n-1) + 3n$$
$$+ 1]P_n(z)x^n] = 0$$

which becomes:
$$f''(x,z) - z[3f'(x,z) + 2xf''(x,z)]$$
$$+ [x^2 f''(x,z) + 3xf'(x,z) + f(x,z)] = 0.$$

Thus,
$$(1 - 2xz + x^2)f'' + (3x - 3z)f' + f = 0.$$

Direct substitution of $f(x,z) = (1 - 2xz + x^2)^{-1/2}$ shows that it satisfies the equation.

(b)Multiply the index shifted equation (like before) by $x^{n+1}/(n+1)!$ with summation n=0 to ∞:
$$\sum_{n=0}^{\infty} \frac{(n+2)P_{n+2}(z)x^{n+1}}{(n+1)!} - \sum_{n+0}^{\infty} \frac{(2n+3)P_{n+1}(z)x^{n+1}}{(n+1)!}$$
$$+ \sum_{n=0}^{\infty} \frac{(n+1)P_n(z)x^{n+1}}{(n+1)!} = 0$$

Pulling a 'd/dx' out front, then a second time for the (n+2) indexed polynomial, then multiply by 'x' and make use of the $g(x,z) = \sum_{n=0}^{\infty} \frac{P_n(z)x^n}{n!}$ substitution:
$$xg'' + (1 - 2zx)g' + (x - z)g = 0.$$

If we now substitute the possible solution $g(x,z) = e^{xz} J_0(x\sqrt{1-z^2})$, where J_0 is just a function at this point (we will see it is the zeroth Bessel function soon) and we get the relation:
$$x\sqrt{1-z^2}J_0''\left(x\sqrt{1-z^2}\right) + J_0'\left(x\sqrt{1-z^2}\right) + x\sqrt{1-z^2}J_0\left(x\sqrt{1-z^2}\right).$$

If we substitute $t = x\sqrt{1-z^2}$, then we have:
$$ty'' + y' + ty = 0,$$

where this is the zeroth order Bessel equation with solution y usually denoted J_0 as already chosen.

Example A.11.
(a) The Bessel functions $J_n(z)$ satisfy the difference equation
$$J_{n+1}(z) - \frac{2n}{z}J_n(z) + J_{n-1}(z) = 0 \qquad (-\infty < n < \infty)$$
with $J_0(0) = 1$ and $J_n(0) = 0$. Define the generating function $f(x,z)$ by

$$f(x,z) = \sum_{n=-\infty}^{\infty} x^n J_n(z).$$

Show that $f(x,z) = exp\left(\frac{z}{2}(x - 1/x)\right)$.

(b) Show that $J_{-n}(z) = J_n(-z) = (-1)^n J_n(z)$.

(c) Show that $1 = J_0(z) + 2\sum_{n=1}^{\infty} J_{2n}(z)$.

Solution

(a) $J_{n+1}(z) - \frac{2n}{z}J_n(z) + J_{n-1}(z) = 0$ is regrouped, using $f(x,z) = \sum_{n=-\infty}^{\infty} x^n J_n(z)$ as:

$$\left(\frac{1}{x}+x\right)f = \frac{2x}{z}f' \quad \rightarrow \quad f(x,z) = exp\left(\frac{z}{2}\left(x-\frac{1}{x}\right)\right)$$

(b) We will use $ex\,p\left(\frac{z}{2}\left(x-\frac{1}{x}\right)\right) = \sum_{n=-\infty}^{\infty} x^n J_n(z)$:

$$\sum_{n=-\infty}^{\infty} x^n J_{-n}(z) = \sum_{n=-\infty}^{\infty} x^{-n} J_n(z) = \sum_{n=-\infty}^{\infty} x^n (-1)^n J_n(z)$$

$$\rightarrow \qquad J_{-n}(z) = (-1)^n J_n(z)$$

Similarly,

$$\sum_{n=-\infty}^{\infty} x^n J_{-n}(z) = \sum_{n=-\infty}^{\infty} y^n J_n(z) = exp\left(\frac{z}{2}\left(y-\frac{1}{y}\right)\right)$$

$$= exp\left(\frac{z}{2}\left(\frac{1}{x} - x\right)\right) = \sum_{n=-\infty}^{\infty} x^n J_n(-z),$$

thus $J_{-n}(z) = J_n(-z)$.

(c)

$$J_0(z) + 2\sum_{n=1}^{\infty} J_{2n}(z) = \sum_{n=-\infty}^{\infty} J_{2n}(z) = \sum_{n=-\infty}^{\infty} x^m J_m(z) \text{ (with } m$$
$$= 2n \text{ and } x = 1).$$

Thus,

$$J_0(z) + 2\sum_{n=1}^{\infty} J_{2n}(z) = exp\left(\frac{z}{2}\left(\frac{1}{1} - 1\right)\right) = 1,$$

thus the result is shown.

Example A.12. Classify all singular points of the following equations (Examine the singularity at infinity as well.):

(a) $x(1-x)y'' + [c - (a+b+1)x]y' - aby = 0$ (the Hypergeometric equation).

(b) $y'' + (h - 2\theta \cos 2x)y = 0$ (the Mathieu equation).

Solution

(a)

$$y'' + \left[\frac{c}{x(1-x)} - \frac{(a+b+1)}{1-x}\right]y' - \frac{ab}{x(1-x)}y = 0.$$

In the neighborhood of the origin we see that x=1 is a regular singular point and x= 0 is an irregular singular point. To examine behavior at infinity let $x = 1/t$:

$$y'' + \left(\frac{(2-c)t + (a+b-1)}{t(t-1)}\right)y' - \frac{ab}{(t^2(t-1))}y = 0.$$

In the neighborhood of the t-origin we see that t=1 is a regular singular point (thus x=1 is a regular singular point) and t= 0 is an irregular singular point (thus x=∞ is an irregular singular point).

(b) $y'' + (h - 2\theta \cos 2x)y = 0$ has no singularities in the neighborhood of the origin. If we substitute $x = 1/t$, then we get:

$$y'' + \frac{2}{t}y' + \frac{(h - 2\theta \cos 2/t)}{t^4}y = 0$$

For this equation we see that t = 0 is an irregular singular point (oscillates as it blows up), thus $x = \infty$ is an irregular singular point.

Example A.13. Using the Frobenius method determine the series expansion for the two solutions of the modified Bessel equation:

$$y'' + \frac{1}{x}y' - \left(a + \frac{v^2}{x^2}\right)y = 0, \qquad with \ \ v = 1.$$

Solution: Left as an exercise.

Example A.14. Find the leading asymptotic behaviors as $x \to +\infty$ of the following equation

a) $y'' = \sqrt{x}\,y$

b) $y'' = \cosh xy'$

Solution

202

(a) Let's start with the substitution: $y = e^s$ → $y' = s'e^s$ → $y'' = s''e^s + (s')^2 e^s$. Thus,

$$s'' + (s')^2 = \sqrt{x}$$

First case: $s'' \ll (s')^2$ → $s' = \pm x^{1/4}$. Since $s'' = \pm(1/4)x^{-3/4}$ we see that this is consistent with $s'' \ll (s')^2$ as $x \to +\infty$.

Second case: $s'' \gg (s')^2$ → $s'' = \sqrt{x}$ → $s' = \left(\frac{2}{3}\right)x^{3/2}$, which is NOT consistent with $s'' \gg (s')^2$ as $x \to +\infty$.

Leading asymptotic behavior is thus $s' = \pm x^{1/4}$ → $s(x) = \pm\frac{4}{5}x^{5/4} + c(x)$. A full solution can be obtained upon solving for c(x):

$$\pm\frac{1}{4}x^{-3/4} + c'' + c'\left(2x^{1/4} + c'\right) = 0.$$

Again using the method of dominant balance, let's try $c'' \ll c'$ → $c = -(1/8)\ln x$, which is consistent. If we try $c' \ll c''$ it is not consistent. Our solution is thus:

$$y(x) = cx^{-1/8} \exp\left(\pm\frac{4}{5}x^{5/4}\right).$$

(b) Use the substitution: $y = e^s$ → $y' = s'e^s$ → $y'' = s''e^s + (s')^2 e^s$ as before. Thus,

$$s'' + (s')^2 = \cosh x \, s'.$$

Suppose $(s')^2 \gg s''$, then $s = \sinh x + c$, and as $x \to \infty$ we have $(\cosh x)^2 \gg \sinh x$, so consistent. If we try $(s')^2 \ll s''$ the result is inconsistent. So, let's try

$$s = \sinh x + c(x)$$

which gives upon substitution:

$$\sinh x + c'' + (\cosh x + 1)c' = 0.$$

Trying dominant balance again, we get $c(x) \sim -\ln(\cosh x)$, thus $s = \sinh x - \ln(\cosh x)$, and:

$$y(x) \sim c\frac{e^{\sinh x}}{\cosh x}.$$

Example A.15. (Bender and Orszag problem 3.45). One way ascertain the asymptotic behavior of certain integrals is to find differential equations that they satisfy and then to perform a local analysis of differential equation. Use this technique to study the behavior of the following integrals

$$\text{a) } y(x) = \int_0^x \exp(l^2)\, dt \ \text{ as } x \to +1$$
$$\text{b) } y(x) = \int_0^\infty \exp(-xt - 1/t)\, dt \ \text{ as } x$$
$$\to 0^+ \text{ and as } x \to +\infty$$

Solution
Left to the reader.

Example A.16. Find the first three terms in the local behavior as $x \to \infty$ of a particular solution to
$$x^3 y'' + y = x^{-4}$$

Solution
Try $y \gg x^3 y''$, thus $y \sim x^{-4}$, which is consistent. So substitute $y(x) = x^{-4} + c(x)$ to get:
$$c'' x^3 + c = -20x^{-3}.$$
Try $c \gg c'' x^3$, thus $c = -20x^{-3}$, which is consistent. So substitute $y(x) = x^{-4} - 20x^{-3} + d(x)$:
$$x^3 d'' + d = 240x^{-2}.$$
Try $d \gg x^3 d''$, thus $d = 240x^{-2}$, which is consistent. So have
$$y(x) = x^{-4} - 20x^{-3} + 240x^{-2} + e(x).$$

Example A.17. (Bender and Orszag 3.55). Find the location of possible stokes line as $z \to \infty$ for the following differential equation
$$y'' = z^{1/3} y$$

Solution:
Local behavior:
$$y(z) \sim cz^{-1/12} \exp\left(\pm(6/7)\, z^{7/6}\right).$$
Leading behavior:
$$e^{\left(\frac{6}{7}\right) z^{7/6}} \quad and \quad e^{-\left(\frac{6}{7}\right) z^{7/6}}.$$
The Stokes lines are the asymptotes as $z \to \infty$ of the curves
$$Re\left\{e^{\left(\frac{6}{7}\right) z^{\frac{7}{6}}} - \left(-e^{-\left(\frac{6}{7}\right) z^{\frac{7}{6}}}\right)\right\} = 0 \to \frac{12}{7} Re\left\{z^{\frac{7}{6}}\right\} = 0 \to e^{i\frac{7}{6}\theta} = 0.$$
Thus, Stokes lines occur for $z = re^{i\theta}$ when $\theta = \pm\frac{3}{7}(2n + 1)\pi$.

Example A.18. Consider the initial value problem

204

$$y' = \frac{y^2}{1 - xy} \quad \text{with} \quad y(0) = 1.$$

(a) Show that about $x = 0$ there is a Taylor series solution of the form:

$$y = \sum_{n=0}^{\infty} A_n x^n$$

where $A_n = \frac{(n+1)^{n-1}}{n!}$.

(b) Show that the solution satisfies

$$y(x) = \exp(xy)$$

and that this equation may be solved iteratively for y as a limit of nested exponentials

$$y(x) = \lim_{n \to \infty} y_n(x)$$

where $y_{n+1}(x) = \exp(xy_n(x))$. Thus, choose $y_0 = 1$, $y_1 = \exp(x)$, $y_2 = \exp(x \exp(x))$, ... Show that the limit exists when $-e \le x \le 1/e$.

Solution
(a) left as an exercise.
(b) left as an exercise.

Example A.19. The differential operator $y' = \cos(\pi xy)$ is too difficult to solve analytically. If solutions plotted for various values of y(0) they are seen to bunch together as x increases. Could this be predicted using asymptotics? Find the possible leading behaviors of solutions as $x \to \infty$. What are the corrections to these leading behaviors?

Solution (partial):
$y' = \cos(\pi xy)$
Let $y(x) = \frac{1}{\pi x} u(x)$ then $u' = \frac{u}{x} + \pi x \cos u$. Now, as $x \to \infty$ we have $u/x \ll \pi x \cos u$. Thus:

$$u' \sim \pi x \cos u \quad \text{or} \quad \frac{du}{\cos u} \sim \pi x dx$$

Since $\ln(\sec u + \tan u) \sim \frac{\pi x^2}{2} + c$ we have

$$\left| 1 + \frac{\sin u}{\cos u} \right| \sim e^{\frac{\pi x^2}{2} + c}.$$

After some regrouping we see:

205

$$u \sim \sin^{-1}\left\{\frac{-1 \pm \exp(\pi x^2 + 2c)}{1 + \exp(\pi x^2 + 2c)}\right\}$$

Thus:

$$u \sim \begin{Bmatrix} \sin^{-1}(-1) \\ \sin^{-1}(1) \end{Bmatrix} \rightarrow \quad u \sim \begin{Bmatrix} \dfrac{-\pi}{2} + 2k\pi \\ \dfrac{\pi}{2} + 2k\pi \end{Bmatrix} \quad for \quad k = 0,1,2\ldots$$

The rest is left as an exercise.

Example A.20. For the equation $y'' = y^2 + e^x$ make the substitutions $y = e^{x/2} u(x)$, $s = e^{x/4}$ and obtain an equation whose solutions for asymptotically large x behave like elliptic functions of s. Deduce that the singularities of y(x) are separated by distance proportional to $e^{-x/4}$ as $x \to \infty$.

Solution

We have: $y'' = y^2 + e^x$; $y = e^{x/2}u(x)$; $s = e^{x/4}$. From which we get

$$y' = e^{x/2}u'(x) + u(x) + \frac{1}{2}e^{x/2}$$

and

$$y'' = e^{x/2}u''(x) + e^{x/2}u'(x) + \frac{1}{4}e^{x/2}u(x)$$

Substituting we get:

$$\frac{d^2u}{ds^2} + \frac{5}{s}\frac{du}{ds} + \frac{4}{s^2}u = 16(u^2 + 1)$$

For $x \to \infty$, $s \to \infty$ and we approximately have:

$$\frac{d^2u}{ds^2} = (u^2 + 1)16.$$

The latter is an autonomous equation that we solve by the following:

$$\left(\frac{d^2u}{ds^2}\right)\frac{du}{ds} = 16[1 + u^2]\frac{du}{ds}$$

and

$$\frac{1}{2}\left[\frac{du}{ds}\right]^2 = 16[u + u^3/3 + c].$$

This becomes: $\pm 4s = \int \dfrac{du}{\sqrt{2u^3/3 + 2u + 2c}}$, which is an elliptic function of s.

The poles for this are separated by period T: $s(x + \Delta) - s(x) \approx T \rightarrow$

$e^{(x+\Delta)/4} - e^{x/4} \approx T \rightarrow e^{\Delta/4} \sim Te^{-x/4}$. Thus, the singularities are separated by distance proportional to $e^{-x/4}$ as $x \rightarrow \infty$.

Example A.21. Show that the leading behavior of an explosive singularity of the Thomas-Fermi equation $y'' = y^{3/2}x^{-1/2}$ is given by:

$$y(x) \sim \frac{400a}{(x-a)^4} \quad as \ x \rightarrow a.$$

Solution
Working with $y'' = y^{3/2}x^{-1/2}$ let's try $y = A(x-a)^b$, in which case we have $y' = Ab(x-a)^{b-1}$ and $y'' = Ab(b-1)(x-a)^{b-2}$. Substituting these we get:

$$b(b-1)(x-a)^{-\frac{1}{2}b-2} = A^{\frac{1}{2}}x^{-\frac{1}{2}}.$$

For this equation to balance asymptotically $(x-a)^{-\frac{1}{2}b-2}$ must be a constant, thus

$$-\frac{1}{2}b - 2 = 0 \quad \rightarrow \quad b = -4.$$

Balancing the constants we then have A=400a, thus we have for solution at leading order:

$$y(x) \sim \frac{400a}{(x-a)^4} \quad as \ x \rightarrow a.$$

B. LIGO Staff ca 1988 (when I was on Staff as a Grad. Stud.) was only ~30 people.

LIGO STAFF, CALTECH
Bridge Lab

	Room	Phone		Room	Phone
Alex Abramovici	358W	4895 446-4169	Pat Lyon	130A	4597
Cynthia Akutagawa	357W	4098 714/594-6948	Boude Moore	31A	4438 792-6406
Bill Althouse	30A	4481 449-6716	Fred Raab	354W	4053 249-6242
Midge Althouse	36A	2975 449-6716	Martin Regehr	360W	2190 568-1910
Fred Asiri	32A	2971 957-5058	Bob Spero	361W	4437 796-0682
Betty Behnke	102E	2129 446-4828	Kip Thorne	128A	4598
Andrej Čadeš	359W	4219 446-2668	Bert Tinker	365W	4610 805/492-5917
Ron Drever	355W	4291 796-0403	Massimo Tinto	358W	4018 449-2007
Ernie Fransgrote	102E	2131 449-5228	Steve Vass	365W	4610 355-9780
Yekta Gürsel	358W	2136 449-9238	Robbie Vogt	101E	3800 794-7823
Jeff Harman	365W	2160 805/495-2354	Steve Winters	354W	- 584-1931
Greg Hiscott	35A	2974 362-7306	Mike Zucker	356W	4017 789-4345
Larry Jones	32A	2970 805/265-9602			

MISC. PHONE NUMBERS

Bridge Lab	365W	4610	Tony Riewe, JPL 144-201		41864
Roof Machine Shop		4894	Rai Weiss, MIT		617/253-3527
Citgrav Computer		449-6081	Susan Merullo, MIT		617/253-4894
CES Lab Control Room		3980	MIT Lab		617/253-4824
CES Lab Computer		3977			
CES Lab, Louie (North End)		3978			
CES Lab, Huey (East End)		3978			
CES Lab, Dewey (South End)		3979	FAX—MIT LIGO Project		617/258-7839
Conference Room	28A	2965	FAX—Caltech LIGO Project		818/304-9834

10/20/88

C. Data Analysis Primer
C.1 Errors add in Quadrature
There is the old experimental/statistical maxim that *"Errors add in Quadrature"*, which is now derived to be true (in most cases) and is due to the propagation of uncertainties. This description will give us an alternate route to the derivation of the sigma of the mean result above as well. So, consider the situation where we measure the quantity of interest indirectly, i.e., we want to measure 'z' but we have x,y,... where z =f(x,y,...). Thus, we have the general relation:

$$\Delta z = \frac{\partial f}{\partial x} \Delta x + \frac{\partial f}{\partial y} \Delta y + \cdots,$$

(C-1)

from which we can square and average over to get:

$$\overline{(\Delta z)^2} = \left(\frac{\partial f}{\partial x}\right)^2 \overline{(\Delta x)^2} + \left(\frac{\partial f}{\partial y}\right)^2 \overline{(\Delta y)^2} + 2\left(\frac{\partial f}{\partial x}\right)\left(\frac{\partial f}{\partial y}\right) \overline{(\Delta x \Delta y)} + \cdots,$$

(C-2)

Upon averaging, the cross terms being linear will have sign cancellation. Thus, rewriting the average of the squared terms as their variance (or std dev squared) notation then clarifies:

$$\sigma_z^{\,2} = \left(\frac{\partial f}{\partial x}\right)^2 \sigma_x^{\,2} + \left(\frac{\partial f}{\partial y}\right)^2 \sigma_y^{\,2} + \cdots.$$

(C-3)

Returning to the case of repeated measurement on iid rv, we have $f = \bar{x}_N$ and this is simply:

$$\sigma_z^{\,2} = (\sigma_x^{\,2} + \sigma_y^{\,2} + \cdots)/N^2.$$

(C-4)

and the addition of error terms is in quadrature. If we use the errors add in quadrature relation we can directly evaluate the sigma of the mean as:

$$\sigma_z = \frac{\sigma}{\sqrt{N}}.$$

(C-5)

C.2 Distributions
Let's now review some of the key distributions that can result. All of the major distributions of interest can be obtained from a maximum entropy evaluation [24]. This takes Maxwell's proposed distribution-based statistical mechanics unification to a new level (Jaynes [68]), and offers greater understanding of the distributional underpinnings of physical systems. Families of distributions are understood to define a manifold (neuromanifold) and this is discussed in [41] and [44]. Some distributions are special in other ways, as is revealed by their ubiquitous appearance.

The Gaussian distribution, in particular, will stand out in this regard. The prior property that errors add in quadrature is the explanation for this as this property underlies how addition of Gaussian noise sources (or repeated measurements) will result in a new total Gaussian (with Gaussian noise). This, in turn is found to generalize to where the repeated measurement is with any background distribution, even one that is changing, will give rise to a total measurement that tends towards being a Gaussian.

The Geometric distribution (emergent via maxent)
Here we talk of the probability of seeing something after k tries when the probability of seeing that event at each try is "p". Suppose we see an event for the first time after k tries, that means the first (k-1) tries were non-events (with probability (1-p) for each try), and the final observation then occurs with probability p, giving rise to the classic formula for the geometric distribution:

$$P(X=k) = (1-p)^{(k-1)}p$$

$$(C-6)$$

As far as normalization, i.e., do all outcomes sum to one, we have:
$$\text{Total Probability} = \Sigma_{k=1} (1-p)^{(k-1)}p = p[1+(1-p)+(1-p)^2+(1-p)^3+\ldots] = p[1/(1-(1-p))]=1.$$

So total probability already sums to one with no further normalization needed. In Fig. C.1 is a geometric distribution for the case where p=0.8:

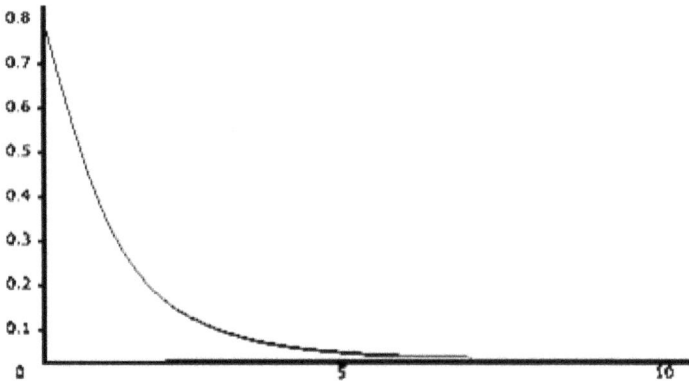

Fig. C.1 The Geometric distribution, $P(X=k) = (1-p)^{(k-1)}p$, with p=0.8.

The Gaussian (aka Normal) distribution (emergent via LLN relation and maxent)

$$N_x(\mu, \sigma^2) = exp(-(x-\mu)^2/(2\sigma^2))/(2\pi\sigma^2)^{(1/2)}$$

For the Normal distribution the normalization is easiest to get via complex integration (so we'll skip that). With mean zero and variance equal one (Fig. C.2) we get:

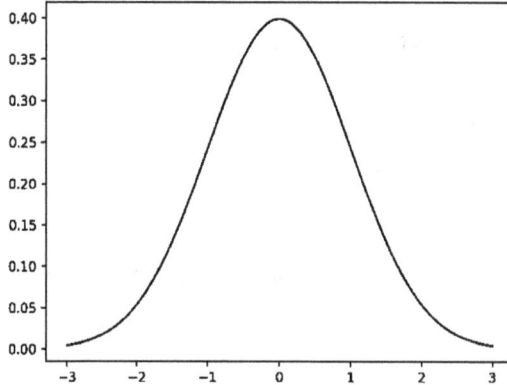

Fig. C.2 The Gaussian distribution, aka Normal, shown with mean zero and variance equal to one: $N_x(\mu, \sigma^2) = N_x(0,1)$.

C.3. Martingales

This section provides a definition of Martingale Processes and shows how many familiar processes are Martingale. When we speak of equilibrium or ergodicity or stationarity we are usually dealing with mathematical objects that are martingales. The properties of equilibrium, a timely convergence of a steady state set of values, e.g., a convergence, is a fundamental property of martingales, thus their frequent appearance in representing processes that arrive at equilibrium. Convergent processes are fundamental to descriptions in statistical mechanics ([44]) as well as to situations (with similar mathematics) in the areas of statistical learning and AI [24].

Martingale Definition[69]

A stochastic process $\{X_n; n=0,1, \ldots\}$ is martingale if, for $n=0,1, \ldots$,

1. $E[|X_n|] < \infty$

2. $E[X_{n+1}|X_0, \ldots, X_n] = X_n$

Def.: Let $\{X_n; n=0,1, \ldots\}$ and $\{Y_n; n=0,1, \ldots\}$ be stochastic processes. We say $\{X_n\}$ is martingale with respect to (w.r.t) $\{Y_n\}$ if, for $n=0,1, \ldots$:

213

1. $E[|X_n|] < \infty$
2. $E[X_{n+1}|Y_0, \ldots, Y_n] = X_n$

Examples of Martingales:

    (a) Sums of independent random variables: $X_n = Y_1 + \ldots + Y_n$.

    (b) Variance of a Sum $X_n = (\sum_{k=1}^{n} Y_k)^2 - n\sigma^2$

    (c) Have induced Martingales with Markov Chains! ….

    (d) For HMM learning, sequences of likelihood ratios are martingale….

The asymptotic equipartition theorem (AEP) and Hoeffding Inequalities (critical in statistical learning [24]) have both been generalized to Martingales.

Induced Martingales with Markov Chains[69]

Let $\{Y_n; n=0,1, \ldots\}$ be a Markov Chain (MC) process with transition probability matrix $P=\|P_{ij}\|$. Let f be a bounded right regular sequence for P:

$f(i)$ is non-negative and $f(i)=\sum_{k=1}^{n} P_{ij}f(j)$. Let $X_n = f(Y_n) \rightarrow E[|X_n|] < \infty$ (since f is bounded). Now have:

$E[X_{n+1}|Y_0, \ldots, Y_n]$

    $= E[f(Y_{n+1})|Y_0, \ldots, Y_n]$

    $= E[f(Y_{n+1})|Y_n]$ (due to MC)

    $= \sum_{k=1}^{n} P_{Y_n,j}f(j)$ (def. of P_{ij} and f)

    $= f(Y_n)$

    $= X_n$

In HMM learning have sequences of likelihood ratios, which is a martingale, proof:

Let $Y_0, Y_1, \ldots$ be iid rv.s and let f_0 and f_1 be probability density functions. A stochastic process of fundamental importance in the theory of testing statistical hypotheses is the sequence of likelihood ratios:

$$X_n = \frac{f_1(Y_0)f_1(Y_1)\ldots f_1(Y_n)}{f_0(Y_0)f_0(Y_1)\ldots f_0(Y_n)}, \quad n = 0,1, \ldots$$

Assume $f_0(y) > 0$ for all y:

$$E[X_{n+1} \mid Y_0, \ldots, Y_n] = E[X_n \left(\frac{f_1(Y_{n+1})}{f_0(Y_{n+1})}\right) \mid Y_0, \ldots, Y_n] = X_n \, E[\frac{f_1(Y_{n+1})}{f_0(Y_{n+1})}]$$

When the common distribution of the Y_k's (used in the 'E' function) has f_0 as its probability density, have:

$$E[\frac{f_1(Y_{n+1})}{f_0(Y_{n+1})}] = 1$$

So, $E[X_{n+1} \mid Y_0, \ldots, Y_n] = X_n$

So likelihood ratios are martingale when the common distribution is f_0.

Random Walk is Martingale [69, pg 238]

Have component-wise proof of random walk for T_Em, both theoretical and computational for a variety of emanators in zero-crossing analysis on Real component in [70]. Since random walk is Martingale (convergence to mean=sqrt(N)) have that Emanation process is Martingale process. In [45] we will see that there may be a unified propagator theory derivative of the choice of emanator theory, where all such theories are martingale. Thus, an argument is provided for why the QFT projection of the emanation process should have processes that are also martingale. Quantum martingales would then relate to the more familiar classical martingales, including their role in classical statistical mechanics ([44]).

Supermartingales and Submartingales[69]

Let $\{X_n; n=0,1, \ldots\}$ and $\{Y_n; n=0,1, \ldots\}$ be stochastic processes. Then $\{X_n\}$ is called a **supermartingale** with respect to $\{Y_n\}$ if, for all n:

    (i) $E[X_n^-] > -\infty$, where $x^- = \min\{x,0\}$

    (ii) $E[X_{n+1} \mid Y_0, \ldots, Y_n] \leq X_n$

    (iii) X_n is a function of $(Y_0, \ldots, Y_n)$ (explicit due to inequality in (ii))

The stochastic process $\{X_n; n=0,1, \ldots\}$ is called a **submartingale** w.r.t $\{Y_n\}$ if, for all n:

    (i) $E[X_n^+] > -\infty$, where $x^+ = \max\{x,0\}$

    (ii) $E[X_{n+1} \mid Y_0, \ldots, Y_n] \geq X_n$

    (iii) X_n is a function of $(Y_0, \ldots, Y_n)$

With Jensen's inequality for convex function φ and conditional expectations have:

$$E[\varphi(X)|Y_0, \ldots, Y_n] \geq \varphi(E[X|Y_0, \ldots, Y_n])$$

So, have means to construct submartingales from martingales (with supermartingales the same aside from a sign flip).

Martingale Convergence Theorems[69]

Under very general conditions, a martingale X_n will converge to a limit random variable X as n increases.

Theorem

(a) Let $\{X_n\}$ be a submartingale satisfying

$$\sup_{n \geq 0} E[|X_n|] < \infty$$

Then there exists a r.v. X_∞ to which $\{X_n\}$ converges with probability one:

$$Prob\left(\lim_{n \to \infty} X_n = X_\infty\right) = 1$$

(b) If $\{X_n\}$ is a martingale and is uniformly integrable, then, in addition to the above, $\{X_n\}$, converges in the mean:

$$\lim_{n \to \infty} E[|X_n - X_\infty|] = 0$$

And $E[X_\infty] = E[X_n]$, for all n.

A sequence is uniformly integral if:

$$\lim_{c \to \infty} \sup_{n \geq 0} E[|X_n|I\{|X_n| > c\}] = 0$$

Where I is the indicator function: 1 if $|X_n| > c$, and 0 otherwise.

'Maximal' Inequalities for Martingales[69]

Chebyshev's inequality applied to a sequence can be 'tightened' to a finer inequality known as the Kolmogorov inequality in terms of the maximum of the sequence. This carries over to Martingales:

Let $\{X_n; n=0,1, \ldots\}$ be iid rvs with $E[X_i]=0$ $\forall$ i and $E[(X_i)^2]=\sigma^2 < \infty$.

Define $S_0 = 0$, $S_n = X_1+\ldots+X_n$, for $n \geq 1$. From Chebyshev's Inequality:

$$\varepsilon^2 Prob(|S_n| > \varepsilon) \leq n\sigma^2, \; \varepsilon > 0$$

A finer inequality is possible:

$$\varepsilon^2 Prob\left(\max_{0 \leq k \leq n} |S_n| > \varepsilon\right) \leq n\sigma^2, \; \varepsilon > 0$$

Known as the Kolmogorov inequality, it can be generalized to provide a maximal inequality on submartingales:

Lemma 1: Let $\{X_n\}$ be a submartingale for which $X_n \geq 0$ for all n. Then for any positive λ:

216

$$\lambda \, Prob \left(\max_{0 \le k \le n} |X_k| > l \right) \le E[X_n]$$

Lemma 2: Let $\{X_n\}$ be a non-negative supermartingale then for any positive λ:

$$\lambda \, Prob \left(\max_{0 \le k \le n} |X_k| > l \right) \le E[X_0]$$

Mean-Square Convergence Theorem for Martingales[69]

Let $\{X_n\}$ be a submartingale w.r.t $\{Y_n\}$ satisfying, for some constant k,

$E[(X_n)^2] \le k < \infty$, for all n. Then $\{X_n\}$ converges as $n \to \infty$ to a limit r.v. X_∞ both with probability one and in mean square:

$$Prob \left(\lim_{n \to \infty} X_n = X_\infty \right) = 1, \quad \text{and} \quad \lim_{n \to \infty} E[|Xn - X_\infty|^2] = 0,$$

Where $E[X_\infty] = E[X_n] = E[X_0]$, for all n.

Martingales w.r.t σ-field formalism

Review of axiomatic probability theory, have three basic elements:

(1) The sample space, a set $\mathbf{\Omega}$ whose elements ω correspond to the possible outcomes of an experiment;

(2) The family of elements, a collection $\mathbf{F}$ of subsets A of $\mathbf{\Omega}$ (the sigma fields). We say that the event A occurs if the outcome ω of the experiment is an element of A;

(3) The probability measure, a function P defined on $\mathbf{F}$ and satisfying:

(i) $0 = P[\varnothing] \le P[A] \le P[\mathbf{\Omega}] = 1$ for $A \in \mathbf{F}$

(ii) $P[A_1 \cup A_2] = P[A_1] + P[A_2] - P[A_1 \cap A_2]$ for $A_i \in \mathbf{F}$

(iii) $P[\cup_{n=1}^{\infty} A_n] = \sum_{n=1}^{\infty} P[An]$ if $A_i \in \mathbf{F}$ are mutually disjoint.

Then, the triple $(\mathbf{\Omega}, \mathbf{F}, P)$ is called a probability space.

Backwards Martingale Definition (w.r.t sigma sub-fields)

Let $\{Z_n\}$ be rv's on a probability space $(\mathbf{\Omega}, \mathbf{F}, P)$ and let $\{G_n; n=0,1, \ldots\}$ be a decreasing sequence of sub sigma-fields of $\mathbf{F}$, viz.,

$$\mathbf{F} \supset \mathbf{F}_n \supset \mathbf{F}_{n+1}, \text{ for all n.}$$

Then $\{Z_n\}$ is called a backward martingale w.r.t. $\{G_n\}$ if for n=0,1, …:

(i) Z_n is G_n-measurable

(ii) $E[|Z_n|] < \infty$, and

(iii) $E[Z_n|G_{n+1}] < Z_{n+1}$

$\{Z_n\}$ is a backwards martingale, iff $X_n = Z_{-n}$, n=0,-1,-2,... forms a martingale w.r.t $F_n = G_{-n}$, n=0,-1,-2,...

Backwards Martingale Convergence Theorem

Let $\{Z_n\}$ be a backwards martingale w.r.t a decreasing sequence of sub sigma-fields $\{G_n\}$. Then:

$$Prob\left(\lim_{n\to\infty} Z_n = Z\right) = 1, \quad and \quad \lim_{n\to\infty} E[|Z - Z_n|] = 0,$$

and $E[Z_n] = E[Z]$, for all n.

Strong Law of Large Numbers Proof

Let $\{X_n; n=1,2, ...\}$ be iid rvs with $E[|X_1|] < \infty$. Let $\mu = E[X_1]$, $S_0 = 0$, and $S_n = X_1+...+X_n$, for $n\geq1$. Let G_n be the sigma field generated by $\{S_n, S_{n+1}, ...\}$. We can derive the strong law of large numbers from the observation that $Z_n = S_n/n$ $(Z_0 = \mu)$, forms a backward martingale w.r.t G_n. Have $E[|Z_n|]< \infty$ and Z_n is G_n-measurable by construction, so just need relation (iii):

$S_n \equiv E[S_n|S_n] = E[S_n|S_n,S_{n+1},...] = E[S_n|G_n] = \sum_{k=1}^{n} E[X_k|G_n] = nE[X_k|G_n],$

with the last equality for $1\leq k\leq n$, thus:

$$Z_n = S_n/n = E[X_k|G_n]$$

So, $E[Z_{n\_1}|G_n] = (n-1)^{-1} E[S_{n-1}|G_n] = (n-1)^{-1} \sum_{k=1}^{n-1} E[X_k|G_n] = Z_n$!!!

Now use backward martingale convergence theorem to show the strong law:

$$Prob\left(\lim_{n\to\infty} \frac{S_n}{n} = \mu\right) = 1$$

C.4. Stationary Processes

A *stationary* process is a stochastic process $\{X(t), t \in T\}$ with the property that for any positive integer 'k', and any points $t_1, ..., t_k$, and h in

T, the joint distribution of $\{X(t_1),...X(t_k)\}$ is the same as the joint distribution of $\{X(t_1+h),...X(t_k+h)\}$.

An ergodic theorem gives conditions under which an average over time

$$\overline{x_n} = \frac{1}{n}(x_1 + \cdots + xn)$$

of a stochastic process will converge as the number n of observed periods becomes large. The strong law of large numbers is one such ergodic theorem.

Stationary processes provide a natural setting for generalization of the law of large numbers since for such processes the mean value is a constant $m=E[X_n]$, independent of time. Just as there are strong and weak laws of large numbers, there are a variety of ergodic theorems.....

Strong Ergodic Theorem [69]

Let $\{X_n; n=0,1, ...\}$ be a strictly stationary process having finite mean $m=E[X_n]$. Let

$$\overline{X_n} = \frac{1}{n}(X_0 + \cdots + X_{n-1})$$

be the sample time average. Then, with probability one, the sequence $\{\overline{X_n}\}$ converges to some limit rv denoted $\overline{X}$:

$$Prob\left(\lim_{n\to\infty} \overline{X_n} = \overline{X}\right) = 1, \quad \text{and} \quad \lim_{n\to\infty} E[|\overline{X} - \overline{X_n}|] = 0,$$

and $E[\overline{X_n}] = E[\overline{X}] = m$.

Asymptotic Equipartition Property (AEP)

$$\lim_{n\to\infty}\left[-\frac{1}{n}\log p(X_0, ..., X_{n-1})\right] = H(\{X_n\})$$

With probability one, provided $\{X_n\}$ is ergodic.

Proof: For $\{X_n\}$ a stationary ergodic finite Markov chain use relation that:

$H(\{X_n\}) = \lim_{k\to\infty} H(Xk|X_1, ..., X_{k-1})$ Or $H(\{X_n\}) = \lim_{l\to\infty} \frac{1}{l} H(X_1, ..., X_l)$

$H(X_n|X_0, ..., X_{n-1}) = -\sum_{i,j} \pi(i)P_{ij} \log P_{ij}$, where $\pi(i)$ is the prior on X_i and P_{ij} is the transition probability to go from X_i to X_j. Thus

$H(\{X_n\}) = -\sum_{i,j} \pi(i)P_{ij} \log P_{ij}$, while,

$-\frac{1}{n}\log p(X_0,...,X_{n-1}) = \frac{1}{n}\sum_{i=0}^{n-2} W_i - \frac{1}{n}\log \pi(X_0)$, where $W_i = -\log P_{i,i+1}$

The ergodic theorem applies:

$$\lim_{n\to\infty}\left[-\frac{1}{n}\log p(X_0,...,X_{n-1})\right] = E[W_0] = -\sum_{i,j}\pi(i)P_{ij}\ \log P_{ij}$$

$$= H(\{X_n\})$$

The general AEP proof uses the backwards martingale convergence theorem instead of the ergodic theorem.

C.5. Sums of random variables
Hoeffding's inequality
Hoeffding's inequality provides an upper bound on the probability that the sum of random variables deviates from its expected value (Wassily Hoeffding, 1963 [71]). It's generalized to martingale differences by Azuma [72] and to functions of random variables $\{X_n\}$ with bounded differences (where function is empirical mean of the sequence of variables: $\bar{X} = \frac{1}{n}(X_1+...+X_n)$ recovers the special case of Hoeffding).

Recall:
Let $X_1,...,X_n$ be independent random variables. Assume that the X_i are almost surely bounded: $P(X_i \in [a_i,b_i])=1$. Define the empirical mean of the sequence of variables as:

$$\bar{X} = \frac{1}{n}(X_1+...+X_n)$$

Hoeffding (1963) proves the following:

$$P(\bar{X}-E[\bar{X}] \geq k) \leq \exp(-\frac{2n^2k^2}{\sum_{i=1}^{n}(b_i-ai)^2})$$

$$P(|\bar{X}-E[\bar{X}]| \geq k) \leq 2\ \exp(-\frac{2n^2k^2}{\sum_{i=1}^{n}(b_i-ai)^2})$$

For each X almost surely bounded have another relation if E(X)=0 known as the Hoeffding Lemma:

$$E[e^{\lambda X}] \leq \exp(\frac{\lambda^2(b-a)^2}{8})$$

The proof begins with showing the Lemma as the hard part.......

Hoeffding Lemma Proof
Since $e^{\lambda X}$ is a convex function, we have

$$e^{\lambda X} \leq \frac{b-X}{b-a}e^{\lambda a} + \frac{X-a}{b-a}e^{\lambda b}, \ \forall\ a \leq x \leq b$$

So,

$E[e^{\lambda X}] \leq E\left[\frac{b-X}{b-a} e^{\lambda a} + \frac{X-a}{b-a} e^{\lambda b}\right] = \frac{b}{b-a} e^{\lambda a} + \frac{-a}{b-a} e^{\lambda b}$ (last is since E[X]=0)

The convexity method involves a line interpolation, let's shift to those parameters with

p = -a/(b-a), and introduce hp = -aλ (so have h = λ(b-a)):

$$\frac{b}{b-a} e^{\lambda a} + \frac{-a}{b-a} e^{\lambda b} = e^{\lambda a}[1-p + p\, e^{\lambda(b-a)}] = e^{-hp}[1-p + p\, e^{h}]$$

$E[e^{\lambda X}] \leq e^{L(h)}$, where L(h) = -hp + ln(1-p+pe^h) $\rightarrow$ L(0) = 0.

L'(h) = -p + pe^h/(1-p+pe^h) $\rightarrow$ L'(0) = 0.

L''(h) = p(1-p)e^h $\rightarrow$ L''(0) = p(1-p).

L$^{(n)}$(h) = p(1-p)e^h > 0

Using Taylor series for L(h):

L(h) = L(0) + hL'(0) + $\frac{1}{2}$h^{2}L''(0) + (more positive terms at higher order in h)

L(h) $\leq \frac{1}{2}$h^2 p(1-p)

Since we have E[X]=0, have p=-a/(b-a) is $\in$ [0,1], so classic logistic function, where the maximum value of p(1-p) on range [0,1] is ¼ (when p=1/2), so:

L(h) $\leq \frac{1}{8}$h^2 and $E[e^{\lambda X}] \leq e^{\frac{1}{8}\lambda^2 (b-a)^2}$

Hoeffding Inequality Proof (for further details, see [71])

Consider Sum on iid X_i , where $S_m = m\bar{X}$ where $\bar{X}$ has m terms in its empirical average:

$P(S_m - E[S_m] \geq k) \leq e^{-tk} E[e^{t(S_m - E[S_m])}]$ (Chernoff Bounding Technique)

$\qquad = \prod_{i=1}^{m} e^{-tk} E[e^{t(X_i - E[X_i])}]$ ({X_n} are iid)

$\qquad \leq \prod_{i=1}^{m} e^{-tk} e^{\frac{1}{8}t^2 (b_i - a_i)^2}$ (Hoeffding Lemma)

$\qquad = e^{-tk} e^{\frac{1}{8}t^2 \sum_{i=1}^{m}(b_i - a_i)^2}$

Have f(t) = -tk+$\frac{1}{8}t^2 \sum_{i=1}^{m}(b_i - a_i)^2$; Choose t=4k/$\sum_{i=1}^{m}(b_i - a_i)^2$ to minimize the upper bound to get:

$$P(S_m - E[S_m] \geq k) \leq e^{-2k^2/\sum_{i=1}^{m}(b_i - a_i)^2}$$
$$P(\bar{X} - E[\bar{X}] \geq k) \leq e^{-2m^2 k^2/\sum_{i=1}^{m}(b_i - a_i)^2}$$

221

Chernoff Bounding Technique:

$P[X \geq k] = P[e^{tX} \geq e^{tk}] \leq e^{-tk} E[e^{tX}]$ (Chernoff uses Markov Inequality on last).

References

[1] Newton, Isaac. "Philosophiæ Naturalis Principia Mathematica. July 5, 1687 (three volumes in Latin). English version: "The Mathematical Principles of Natural Philosophy", Encyclopædia Britannica, London. (1687).

[2] Leibniz, Gottfried Wilhelm Freiherr von; Gerhardt, Carl Immanuel (trans.) (1920). The Early Mathematical Manuscripts of Leibniz. Open Court Publishing. p. 93. Retrieved 10 November 2013..

[3] Dirk Jan Struik, A Source Book in Mathematics (1969) pp. 282–28.

[4] Leibniz, Gottfried Wilhelm. Supplementum geometriae dimensoriae, seu generalissima omnium tetragonismorum effectio per motum: similiterque multiplex constructio lineae ex data tangentium conditione, Acta Euriditorum (Sep. 1693) pp. 385–392.

[5] Euler, Leonhard. Mechanica sive motus scientia analytice exposita; 1736.

[6] Laplace, P S (1774), "Mémoires de Mathématique et de Physique, Tome Sixième" [Memoir on the probability of causes of events.], Statistical Science, 1 (3): 366–367.

[7] D'Alembert, Jean Le Rond (1743). Traité de dynamique .

[8] Lagrange, J. L. , Mécanique analytique, Vol. 1 (1788), Vol. 2 (1789). Expanded republished Vol. 1 1811 and Vol. 2 1815.

[9] Lagrange, J. L. (1997). Analytical mechanics. Vol. 1 (2d ed.). English translation of the 1811 edition.

[10] William R. Hamilton. On a General Method in Dynamics; by which the Study of the Motions of all free Systems of attracting or repelling Points is reduced to the Search and Differentiation of one central Relation, or characteristic Function. Philosophical Transactions of the Royal Society (part II for 1834, pp. 247-308).

[11] William R. Hamilton. Second Essay on a General Method in Dynamics'. This was published in the Philosophical Transactions of the Royal Society (part I for 1835, pp. 95-144).

[12] Hamilton, W. (1833). "On a General Method of Expressing the Paths of Light, and of the Planets, by the Coefficients of a Characteristic Function" (PDF). Dublin University Review: 795–826.

[13] Hamilton, W. (1834). "On the Application to Dynamics of a General Mathematical Method previously Applied to Optics" (PDF). British Association Report: 513–518.

[14] W.R. Hamilton(1844 to 1850) On quaternions or a new system of imaginaries in algebra, Philosophical Magazine,

[15] Simon L. Altmann (1989). "Hamilton, Rodrigues and the quaternion scandal". Mathematics Magazine. Vol. 62, no. 5. pp. 291–308.

[16] Werner Heisenberg (1925). "Über quantentheoretische Umdeutung kinematischer und mechanischer Beziehungen". Zeitschrift für Physik (in German). 33 (1): 879–893. ("Quantum theoretical re-interpretation of kinematic and mechanical relations")

[17] Schrödinger, E. (1926). "An Undulatory Theory of the Mechanics of Atoms and Molecules" (PDF). Physical Review. 28 (6): 1049–1070.

[18] Dirac, Paul Adrien Maurice (1930). The Principles of Quantum Mechanics. Oxford: Clarendon Press.

[19] Feigenbaum, M. J. (1976). "Universality in complex discrete dynamics" (PDF). Los Alamos Theoretical Division Annual Report 1975–1976.

[20] Morse, Marston (1934). The Calculus of Variations in the Large. American Mathematical Society Colloquium Publication. Vol. 18. New York.

[21] Milnor, John (1963). Morse Theory. Princeton University Press. ISBN 0-691-08008-9.

[22] Fizeau, H. (1851). "Sur les hypothèses relatives à l'éther lumineux". Comptes Rendus. 33: 349–355.

[23] Shankland, R. S. (1963). "Conversations with Albert Einstein". American Journal of Physics. 31 (1): 47–57.

[24] Winters-Hilt, S. Informatics and Machine Learning: from Martingales to Metaheuristics. (2021) Wiley.

[25] Goldstein, Herbert (1980). Classical Mechanics (2nd ed.). Addison-Wesley.

[26] Neother, E. (1918). "Invariante Variationsprobleme". Nachrichten von der Gesellschaft der Wissenschaften zu Göttingen.Mathematisch-Physikalische Klasse.1918: 235-257.

[27] Landau, Lev D.; Lifshitz, Evgeny M. (1969). Mechanics. Vol. 1 (2nd ed.). Pergamon Press.

[28] Percival, I.C. and D. Richards. Introduction to Dynamics. (1983) Cambridge University Press.

[29] Fetter, A.L and J.D Walecka, Theoretical Mechanics of Particles and Continua, Dover (2003).

[30] Kapitza, P.L. "Dynamic stability of the pendulum with vibrating suspension point," Sov. Phys. JETP 21 (5), 588–597 (1951) (in Russian).

[31] Lyapunov, A.M. The general problem of the stability of motion. 1892. Kharkiv Mathematical Society, Kharkiv, 251p. (in Russian).

[32] Arnold, V.I. Ordinary Differential Equations. MIT Press. (1978).

[33] Longair, M.S. Theoretical Concepts in Physics: An Alternative View of Theoretical Reasoning in Physics. Cambridge University Press. 2nd edition: 2003.

[34] Baker, G.L and J. Gollub. Chaoric Dynamics: An Introduction. Cambridge University Press. 1990.

[35] Mandelbrot, Benoît (1982). The Fractal Geometry of Nature. W H Freeman & Co.

[36] P.J. Myrberg. Iteration der rellen Polynome zweiten Grades. III, Annales Acad. Sci Fenn A, U 336 (1963) n.3, 1-18, MR 27.

[37] Arnold, Vladimir I. (1989). Mathematical Methods of Classical Mechanics (2nd ed.). New York: Springer.

[38] Woodhouse, N.M.J. Introduction to Analytical Dynamics. Springer, 2nd Edition. 2009.

[39] Bender, C.M. and S.A. Orszag. Advanced Mathematical Methods for Scientists and Engineers: Asymptotic Methods and Perturbation Theory. Springer. 1999.

[40] Winters-Hilt, S. The Dynamics of Fields, Fluids, and Gauges. (Physics Series: "Physics from Maximal Information Emanation" Book 2.)

[41] Winters-Hilt, S. The Dynamics of Manifolds. (Physics Series: "Physics from Maximal Information Emanation" Book 3.)

[42] Winters-Hilt, S. Quantum Mechanics, Path Integrals, and Algebraic Reality. (Physics Series: "Physics from Maximal Information Emanation" Book 4.)

[43] Winters-Hilt, S. Quantum Field Theory and the Standard Model. (Physics Series: "Physics from Maximal Information Emanation" Book 5.)

[44] Winters-Hilt, S. Thermal & Statistical Mechanics, and Black Hole Thermodynamics. (Physics Series: "Physics from Maximal Information Emanation" Book 6.)

[45] Winters-Hilt, S. Emanation, Emergence, and Eucatastrophe. (Physics Series: "Physics from Maximal Information Emanation" Book 7.)

[46] Winters-Hilt, S. Classical Mechanics and Chaos. (Physics Series: "Physics from Maximal Information Emanation" Book 1.)

[47] Winters-Hilt, S. Data analytics, Bioinformatics, and Machine Learning. 2019.

[48] Feynman, R.P. and A.R. Hibbs. Quantum Mechanics and Path Integrals. McGraw-Hill College. 1965.

[49] Landau, L.D.; Lifshitz, E.M. (1935). "Theory of the dispersion of magnetic permeability in ferromagnetic bodies". Phys. Z. Sowjetunion. 8, 153.

[50] Landau, Lev D.; Lifshitz, Evgeny M. (1980). Statistical Physics. Vol. 5 (3rd ed.). Butterworth-Heinemann.

[51] Braginskii, V. B. Measurement of weak forces in physics experiments. (1977). University of Chicago Press.

[52] Drever, R. W. P.; Hall, J. L.; Kowalski, F. V.; Hough, J.; Ford, G. M.; Munley, A. J.; Ward, H. (June 1983). "Laser phase and frequency stabilization using an optical resonator" (PDF). Applied Physics B. 31 (2): 97–105.

[53] Bunimovich, V.I. Fluctuational processes in radioreceivers. Gostekhizdat, USSR. 1950.

[54] Stratonovich, R.L. Selected problems in the theory of fluctuations in radiotechnology. Soviet Radio, USSR.

[55] Papoulis, Athanasios; Pillai, S. Unnikrishna (2002). Probability, Random Variables and Stochastic Processes (4th ed.). Boston: McGraw Hill.

[56] Reed, M, and Simon, B. Methods of modern mathematical physics. III. Scattering theory. Elsevier, 1979.

[57] Rutherford, E. (1911). "LXXIX. The scattering of α and β particles by matter and the structure of the atom". The London, Edinburgh, and Dublin Philosophical Magazine and Journal of Science. 21 (125): 669–688.

[58] Sommerfeld, Arnold (1916). "Zur Quantentheorie der Spektrallinien". Annalen der Physik. 4 (51): 51–52.

[59] Hibbeler, R. Engineering Mechanics: Dynamics. 14th Edition. 2015.

[60] Hibbeler, R. Engineering Mechanics: Statics and Dynamics. 14th Edition. 2015.

[61] Layek, G.C. An Introduction to Dynamical Systems and Chaos 1st ed. 2015. Springer.

[62] Lemons, D.S. A Student's Guide to Dimensional Analysis. Cambridge University Press. 1st edition: 2017.

[63] Langhaar, H.L. Dimensional Analysis and Theory of Models, Wiley 1951.

[64] Feynman, R. P. (1948). The Character of Physical Law. MIT Press (1967).

[65] Ince, E. L. Ordinary Differential Equations. Dover 1956.

[66] Abromowitz, M. and I.A. Stegun. Handbook of Mathematical Functions. Dover 1965.

[67] Fuchs, L.I. On the theory of linear differential equations with variable coefficients. 1866.

[68] Jaynes, E. T. Probability Theory: The Logic of Science. Cambridge University Press, (2003).

[69] Karlin, S. and H.M. Taylor. A First Course in Stochastic Processes 2nd Ed. Academic Press. 1975.

[70] Winters-Hilt, S. Unified Propagator Theory and a non-experimental derivation for the fine-structure constant. Advanced Studies in Theoretical Physics, Vol. 12, 2018, no. 5, 243-255.

[71] Wassily Hoeffding (1963) Probability inequalities for sums of bounded random variables, *Journal of the American Statistical Association*, 58 (301), 13–30.

[72] Azuma, K. (1967). "Weighted Sums of Certain Dependent Random Variables" (PDF). *Tôhoku Mathematical Journal*. **19** (3): 357–367.

[73] Compton, Arthur H. (May 1923). "A Quantum Theory of the Scattering of X-Rays by Light Elements". Physical Review. 21 (5): 483–502.

[74] Mason and Woodhouse. "Relativity and Electromagnetism" (PDF). Retrieved 20 February 2021.

[75] Merzbach, Uta C.; Boyer, Carl B. (2011), *A History of Mathematics* (3rd ed.), John Wiley & Sons.

[76] Robinson, Abraham (1963), Introduction to model theory and to the metamathematics of algebra, Amsterdam: North-Holland, ISBN 978-0-7204-2222-1, MR 0153570

[77] Robinson, Abraham (1966), Non-standard analysis, Princeton Landmarks in Mathematics (2nd ed.), Princeton University Press, ISBN 978-0-691-04490-3, MR 0205854

[78] R. D. Richtmyer (1978), *Principles of Advanced Mathematical Physics* Vol. 1 & 2, Springer-Verlag, New York.

[79] Tufillaro, N., T. Abbott and D. Griffiths. Swinging Atwood's Machine. American Journal of Physics, 52, 895–903, 1984.

[80] https://en.wikipedia.org/wiki/Logistic_map

[81] Winters-Hilt S. Topics in Quantum Gravity and Quantum field Theory in Curved Spacetime. UWM PhD Dissertation, 1997.

[82] Winters-Hilt S, I. H. Redmount, and L. Parker, "Physical distinction among alternative vacuum states in flat spacetime geometries," Phys. Rev. D 60, 124017 (1999).

[83] Friedman J. L., J. Louko, and S. Winters-Hilt, "Reduced Phase space formalism for spherically symmetric geometry with a massive dust shell," Phys. Rev. D 56, 7674-7691 (1997).

[84] Louko J and S. Winters-Hilt, "Hamiltonian thermodynamics of the Reissner-Nordstrom-anti de Sitter black hole," Phys. Rev. D 54, 2647-2663 (1996).

[85] Louko J, J. Z. Simon, and S. Winters-Hilt, "Hamiltonian thermodynamics of a Lovelock black hole," Phys. Rev. D 55, 3525-3535 (1997).

[86] Amari, S. and H. Nagaoka. Methods of Information Geometry. Oxford University Press. 2000.

[87] Winters-Hilt, S. Feynman-Cayley Path Integrals select Chiral Bi-Sedenions with 10-dimensional space-time propagation. Advanced Studies in Theoretical Physics, Vol. 9, 2015, no. 14, 667-683.

[88] Winters-Hilt, S. The 22 letters of reality: chiral bisedenion properties for maximal information propagation. Advanced Studies in Theoretical Physics, Vol. 12, 2018, no. 7, 301-318.

[89] Winters-Hilt, S. Fiat Numero: Trigintaduonion Emanation Theory and its Relation to the Fine-Structure Constant α, the Feigenbaum Constant C_∞, and π. Advanced Studies in Theoretical Physics, Vol. 15, 2021, no. 2, 71-98.

[90] Winters-Hilt, S. Chiral Trigintaduonion Emanation Leads to the Standard Model of Particle Physics and to Quantum Matter. Advanced Studies in Theoretical Physics, Vol. 16, 2022, no. 3, 83-113.

[91] Robert L. Devaney. An Introduction to Chaotic Dynamical Systems. Addison -Wesley.

[92] Landau, Lev D.; Lifshitz, Evgeny M. (1971). *The Classical Theory of Fields*. Vol. 2 (3rd ed.). Pergamon Press.

[93] Penrose, Roger (1965), "Gravitational collapse and space-time singularities", Phys. Rev. Lett., 14 (3): 57.

[94] Hawking, Stephen & Ellis, G. F. R. (1973). The Large Scale Structure of Space-Time. Cambridge: Cambridge University Press.

[95] Peebles, P. J. E. (1980). Large-Scale Structure of the Universe. Princeton University Press.

[96] B. Abi et al. Measurement of the Positive Muon Anomalous Magnetic Moment to 0.46 ppm
Phys. Rev. Lett. 126, 141801 (2021).

[97] Einstein, A. "On a heuristic point of view concerning the production and transformation of light" (Ann. Phys., Lpz 17 132-148)

[98] Balmer, J. J. (1885). "Notiz über die Spectrallinien des Wasserstoffs" [Note on the spectral lines of hydrogen]. Annalen der Physik und Chemie. 3rd series (in German). 25: 80–87.

[99] Bohr, N. (July 1913). "I. On the constitution of atoms and molecules". The London, Edinburgh, and Dublin Philosophical Magazine

and Journal of Science. 26 (151): 1–
25. doi:10.1080/14786441308634955.
[100] Bohr, N. (September 1913). "XXXVII. On the constitution of atoms and molecules". The London, Edinburgh, and Dublin Philosophical Magazine and Journal of Science. 26 (153): 476–
502. Bibcode:1913PMag...26..476B. doi:10.1080/14786441308634993.
[101] Bohr, N. (1 November 1913). "LXXIII. On the constitution of atoms and molecules". The London, Edinburgh, and Dublin Philosophical Magazine and Journal of Science. 26 (155): 857–
875. doi:10.1080/14786441308635031.
[102] Bohr, N. (October 1913). "The Spectra of Helium and Hydrogen". Nature. 92 (2295): 231–232.
[103] Max Planck. On the Law of Distribution of Energy in the Normal Spectrum. Annalen der Physik vol. 4, p. 553 ff (1901)
[104] Arthur H. Compton. Secondary radiations produced by x-rays. Bulletin of the National Research Council., no. 20 (v. 4, pt. 2) Oct. 1922.
[105] Davisson, C. J.; Germer, L. H. (1928). "Reflection of Electrons by a Crystal of Nickel". Proceedings of the National Academy of Sciences of the United States of America. 14 (4): 317–322.
[106] Michael Eckert. How Sommerfeld extended Bohr's model of the atom (1913–1916). The European Physical Journal H.
[107] Max Born; J. Robert Oppenheimer (1927). "Zur Quantentheorie der Molekeln" [On the Quantum Theory of Molecules]. Annalen der Physik (in German). 389 (20): 457–484.
[108] Dirac, P. A. M. (1928). "The Quantum Theory of the Electron" (PDF). Proceedings of the Royal Society A: Mathematical, Physical and Engineering Sciences. 117 (778): 610–624.
[109] Dirac, Paul A. M. (1933). "The Lagrangian in Quantum Mechanics" (PDF). Physikalische Zeitschrift der Sowjetunion. 3: 64–72.
[110] Feynman, Richard P. (1942). The Principle of Least Action in Quantum Mechanics (PDF) (PhD). Princeton University.
[111] Feynman, Richard P. (1948). "Space-time approach to non-relativistic quantum mechanics". Reviews of Modern Physics. 20 (2): 367–387.
[112] Erdeyli, A. Asymptotic Expansions. 1956 Dover.
[113] Erdeyli, A. Asymptotic Expansions of differential equations with turning points. Review of the Literature. Technical Report 1, Contract Nonr-220(11). Reference no. NR 043-121. Department of Mathematics, California Institute of Technology, 1953.
[114] Carrier, G.F, M. Crook and C.E. Pearson. Functions of a complex variable. 1983 Hod Books.

[115] Van Vleck, J. H. (1928). "The correspondence principle in the statistical interpretation of quantum mechanics". Proceedings of the National Academy of Sciences of the United States of America. 14 (2): 178–188.

[116] Chaichian, M.; Demichev, A. P. (2001). "Introduction". Path Integrals in Physics Volume 1: Stochastic Process & Quantum Mechanics. Taylor & Francis. p. 1ff. ISBN 978-0-7503-0801-4.

[117] Vinokur, V. M. (2015-02-27). "Dynamic Vortex Mott Transition"

[118] Hawking, S. W. (1974-03-01). Black hole explosions? Nature. 248 (5443): 30–31.

[119] Birrell, N.D. and Davies, P.C.W. (1982) Quantum Fields in Curved Space. Cambridge Monographs on Mathematical Physics. Cambridge University Press, Cambridge.

[120] Maldacena, Juan (1998). "The Large N limit of superconformal field theories and supergravity". Advances in Theoretical and Mathematical Physics. 2 (4): 231–252.

[121] Witten, Edward (1998). "Anti-de Sitter space and holography". Advances in Theoretical and Mathematical Physics. 2 (2): 253–291.

[122] Caves, Carlton M.; Fuchs, Christopher A.; Schack, Ruediger (2002-08-20). "Unknown quantum states: The quantum de Finetti representation". Journal of Mathematical Physics. 43 (9): 4537–4559.

[123] Jackson, J.D. Classical Electrodynamics, 2nd Edition. Wiley 1975.

[124] Lorentz, Hendrik Antoon (1899), "Simplified Theory of Electrical and Optical Phenomena in Moving Systems" , *Proceedings of the Royal Netherlands Academy of Arts and Sciences*, **1**: 427–442.

[125] Misner, Charles W., Thorne, K. S., & Wheeler, J. A. Gravitation. Princeton University Press, 2017. ISBN: 9780691177793.

[126] Penrose, R., W. Rindler (1984) Volume 1: Two-Spinor Calculus and Relativistic Fields, Cambridge University Press, United Kingdom.

[127] Tolkien, J.R.R. (1990). *The Monsters and the Critics and Other Essays*. London: HarperCollinsPublishers.

Index

cylinder, 74, 171

248